Japanese Abacus For Kids

A step-by-step guide to addition and subtraction using the Japanese abacus (Soroban).

Author:
Paul Green

2

First published 2016

Copyright @ 2016, Paul Green. All rights reserved.

No part of this publication may be reproduced in any material form (including photocopying or storing in any medium by electronic means and whether or not transiently or incidentally to some other use of this publication) without the written permission of the copyright holder.

ISBN-13: 978-1541048645
ISBN-10: 1541048644

PREFACE

The Japanese abacus has been used as a calculation tool for generations and can still be seen in use in Japan today. Children are still taught to use this instrument in schools today. It is widely available, cheap to buy and fun to use.

This book will teach children the skills required to use the abacus effectively, once learnt and practised these skills will stay with them throughout their lives. A useful and impressive skill that would be an asset for anyone.

Important

I'll help you learn to use the Abacus.

CONTENTS

INTRODUCTION

Nice to know

▷ The Japanese abacus is also called the Soroban.

▷ This is abacus written in Japanese そろばん

▷ The Japanese abacus is mostly used for adding and subtracting numbers.

▷ The Japanese abacus has a wooden frame and five beads per column, one bead above and four beads below.

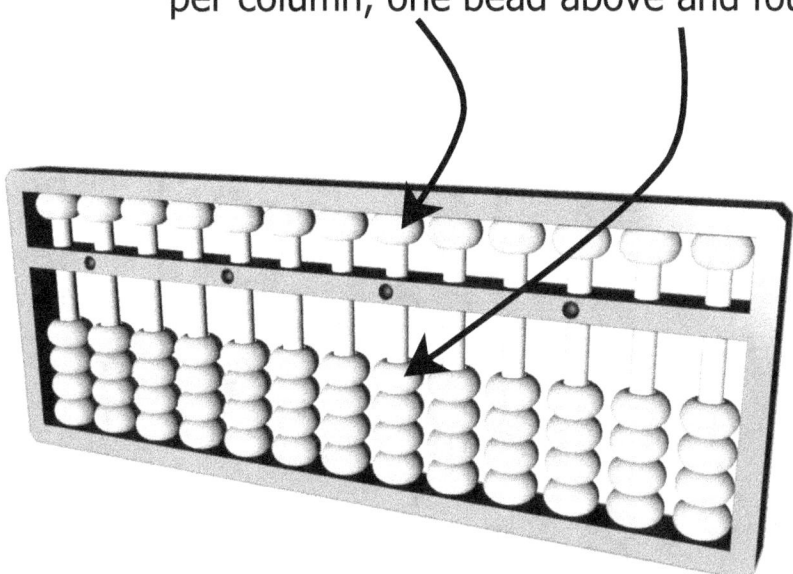

THE PARTS OF THE JAPANESE ABACUS

1. A wooden frame.

2. A beam, to push the beads up against and away from.

3. Dots on the beam.

4. Rods, to slide the beads up and down on.

5. 1 bead above the beam.

6. 4 beads below the beam.

7. A Column is one rod and the 5 beads on that rod. There are 13 columns on this abacus.

8. Lower deck. All beads that are below the beam are in the lower deck.

9. Upper deck. All beads that are above the beam are in the upper deck.

Tip Abaci have different amounts of rods. Usually 13 rods but some have less and some have more.

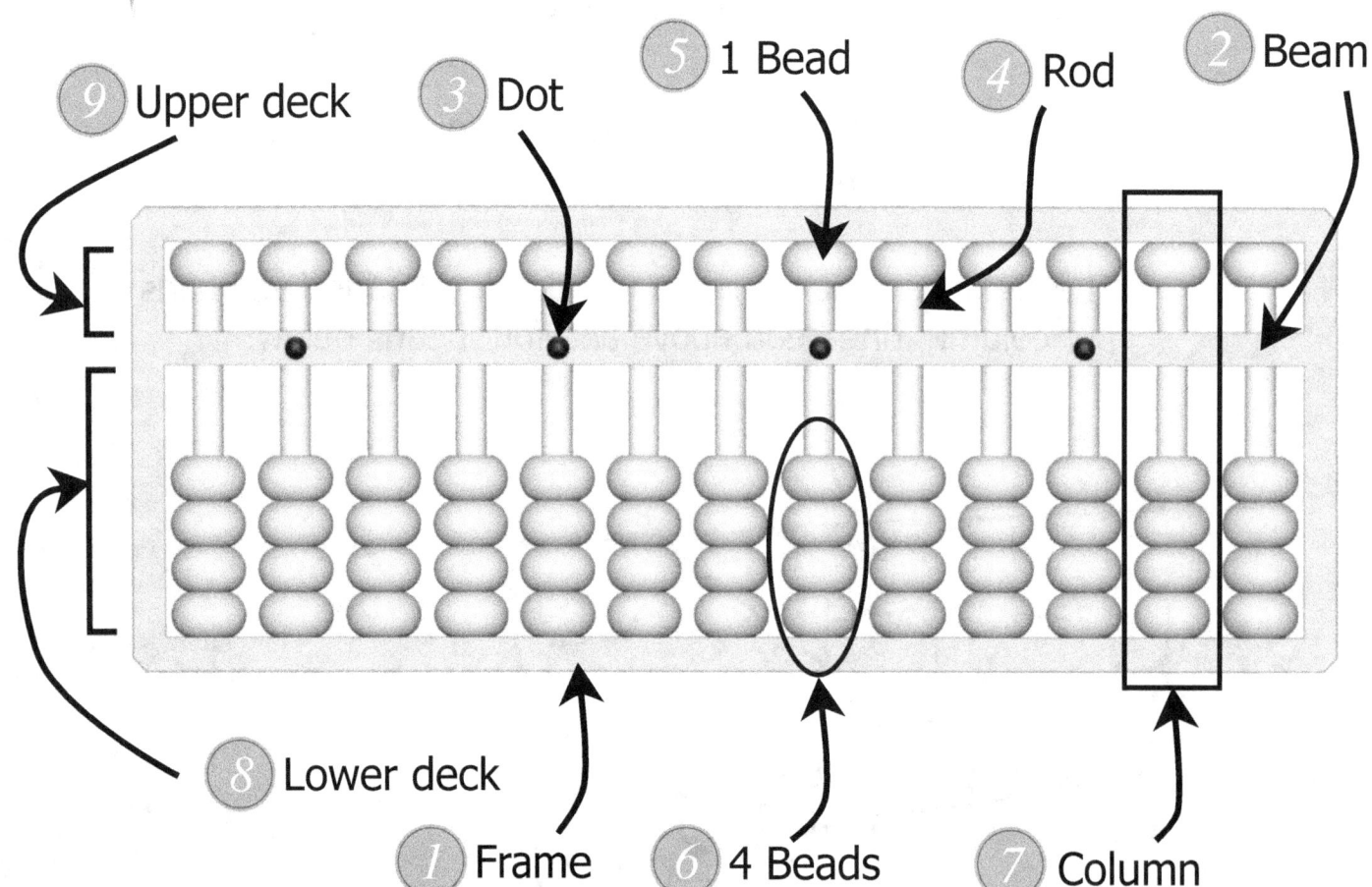

PUTTING YOUR NUMBERS IN THE CORRECT COLUMN

We need to know which abacus column to use to place each digit.

Let's look at the following 4 digit number **4213**

- The number 3 is on the 'Ones' column
- The number 1 is on the 'Tens' column
- The number 2 is on the 'Hundreds' column
- The number 4 is on the 'Thousands' column

4213

This is how it would look on the abacus.

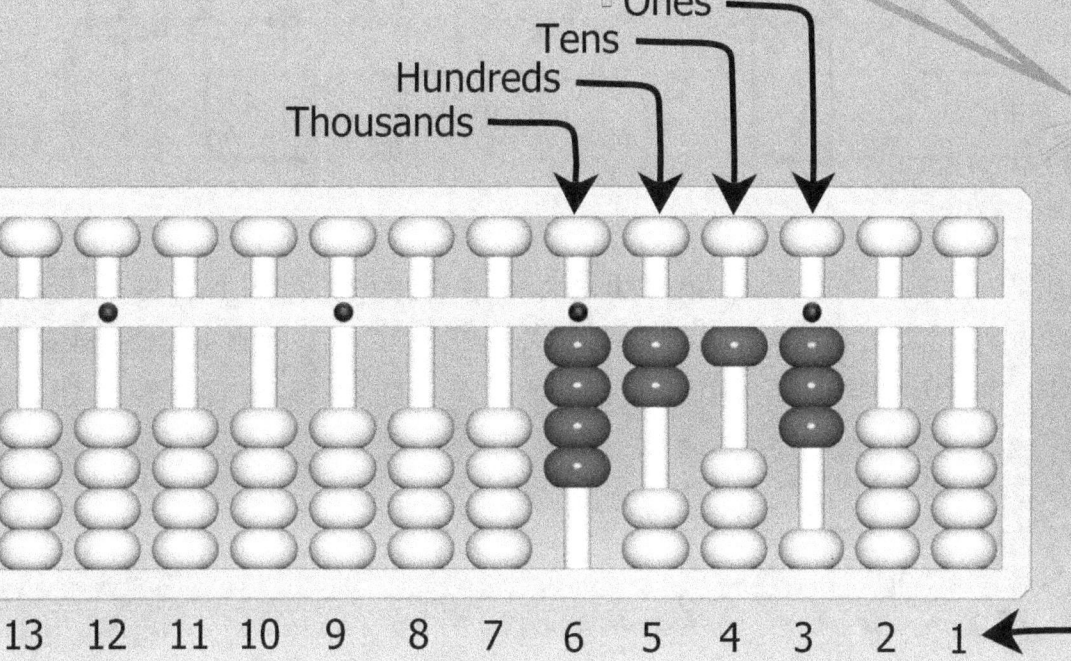

Look how we start with the 'Ones' digit (3 in this example) on column 3 (where the first dot on the beam is) and not on columns 1 and 2. *Don't worry about this, we will learn why later.*

WHAT AMOUNT IS EACH COLUMN WORTH?

The picture below shows the values of each column on the abacus.

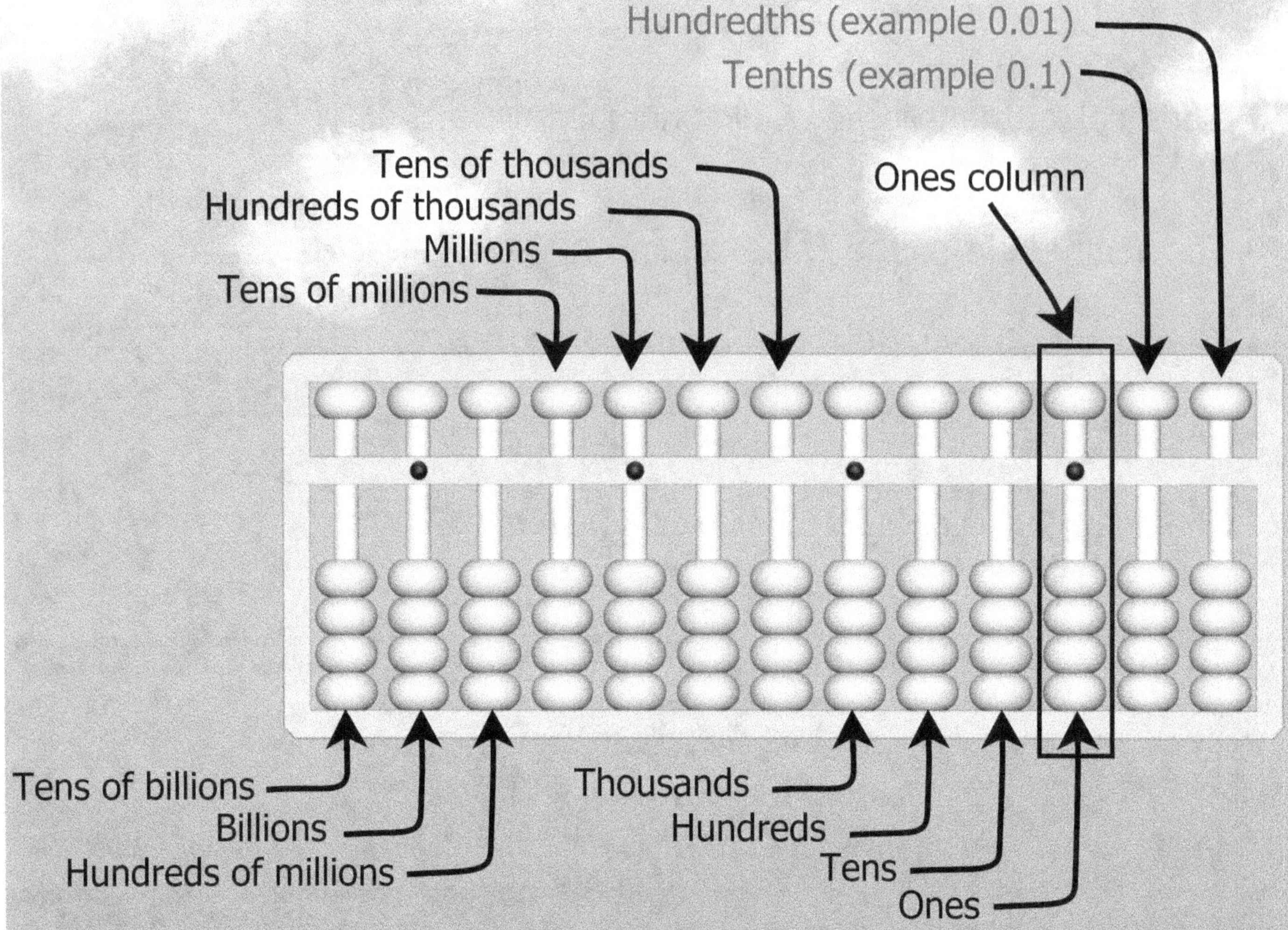

Hundredths (example 0.01)

Tenths (example 0.1)

Tens of thousands

Hundreds of thousands

Millions

Tens of millions

Ones column

Tens of billions

Billions

Hundreds of millions

Thousands

Hundreds

Tens

Ones

Tip Notice how each column value keeps getting TEN times BIGGER on the left of the ones column and keeps getting TEN times smaller on the right of the ones column.

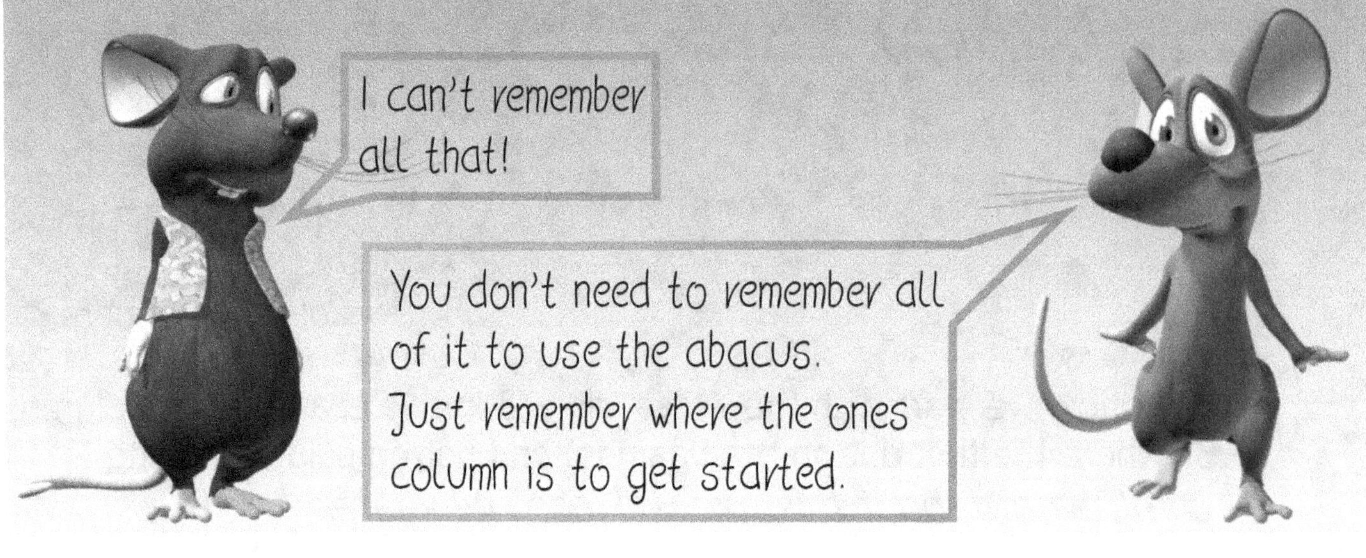

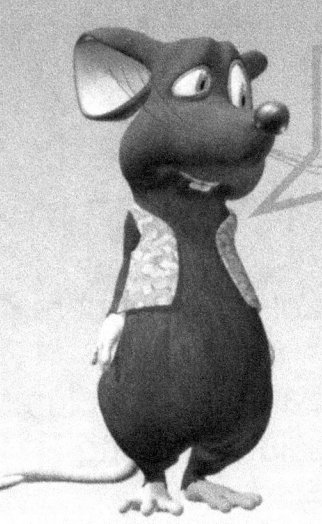

I can't remember all that!

You don't need to remember all of it to use the abacus. Just remember where the ones column is to get started.

ABACUS BASICS

HOW TO MOVE BEADS TO MAKE A NUMBER ON THE ABACUS

Important!

- When we move a bead towards the beam this is called **'Register a bead'.**
- When we move a bead away from the beam this is called **'Unregister a bead'.**
- The bead above the beam is worth **5.**
- The beads below the beam are worth 1.
- We read the result on the abacus by looking only at the beads that are **pushed against the beam.**

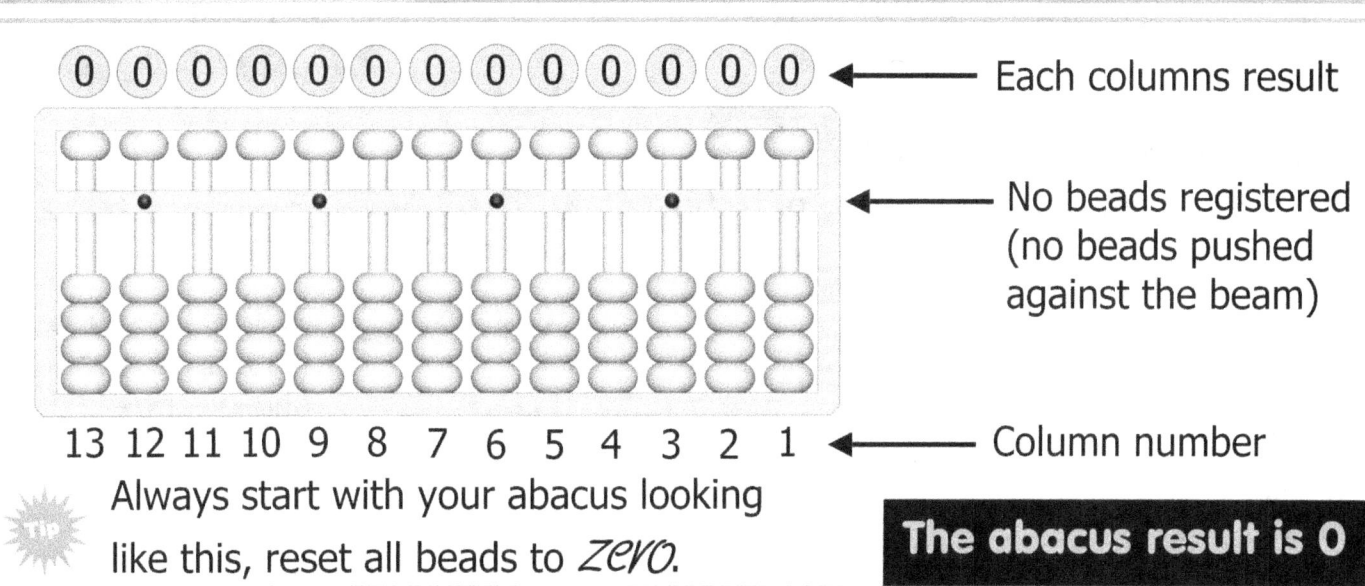

Each columns result

No beads registered (no beads pushed against the beam)

Column number

Always start with your abacus looking like this, reset all beads to *zero*.

The abacus result is 0

Let's put the number 1 on the abacus

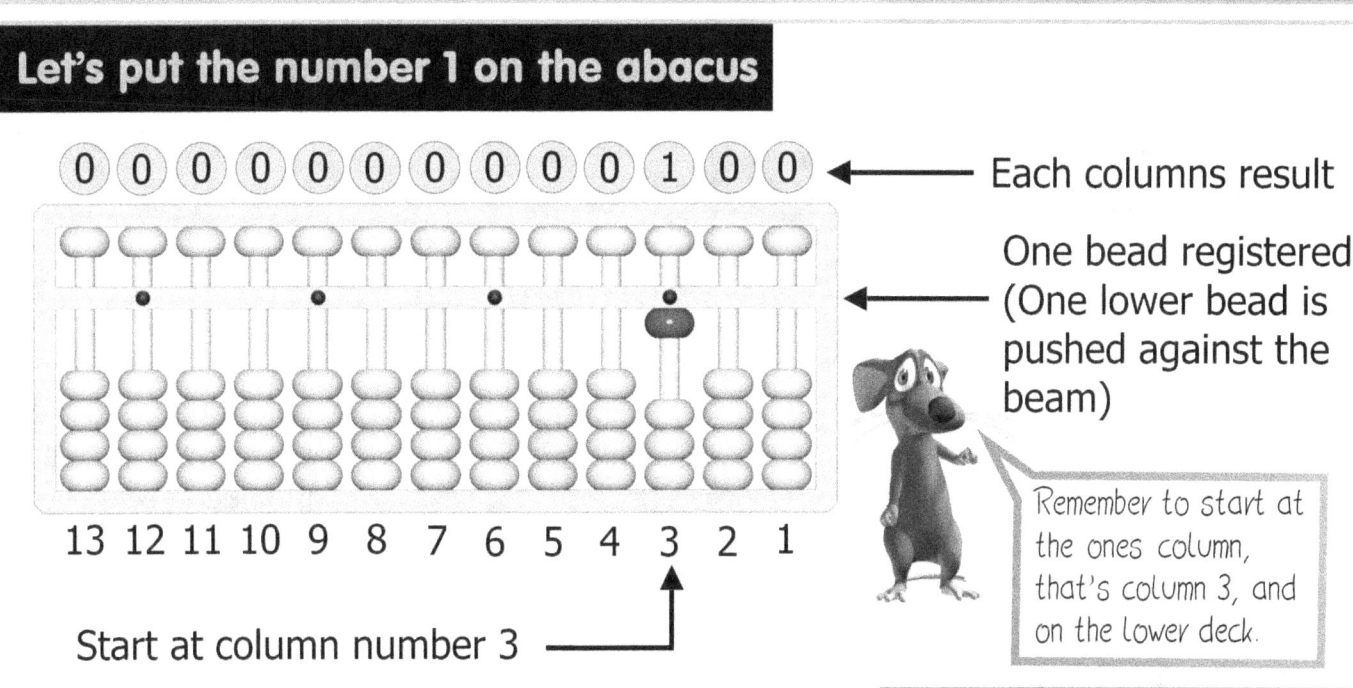

Each columns result

One bead registered (One lower bead is pushed against the beam)

Remember to start at the ones column, that's column 3, and on the lower deck.

Start at column number 3

The abacus result is 1

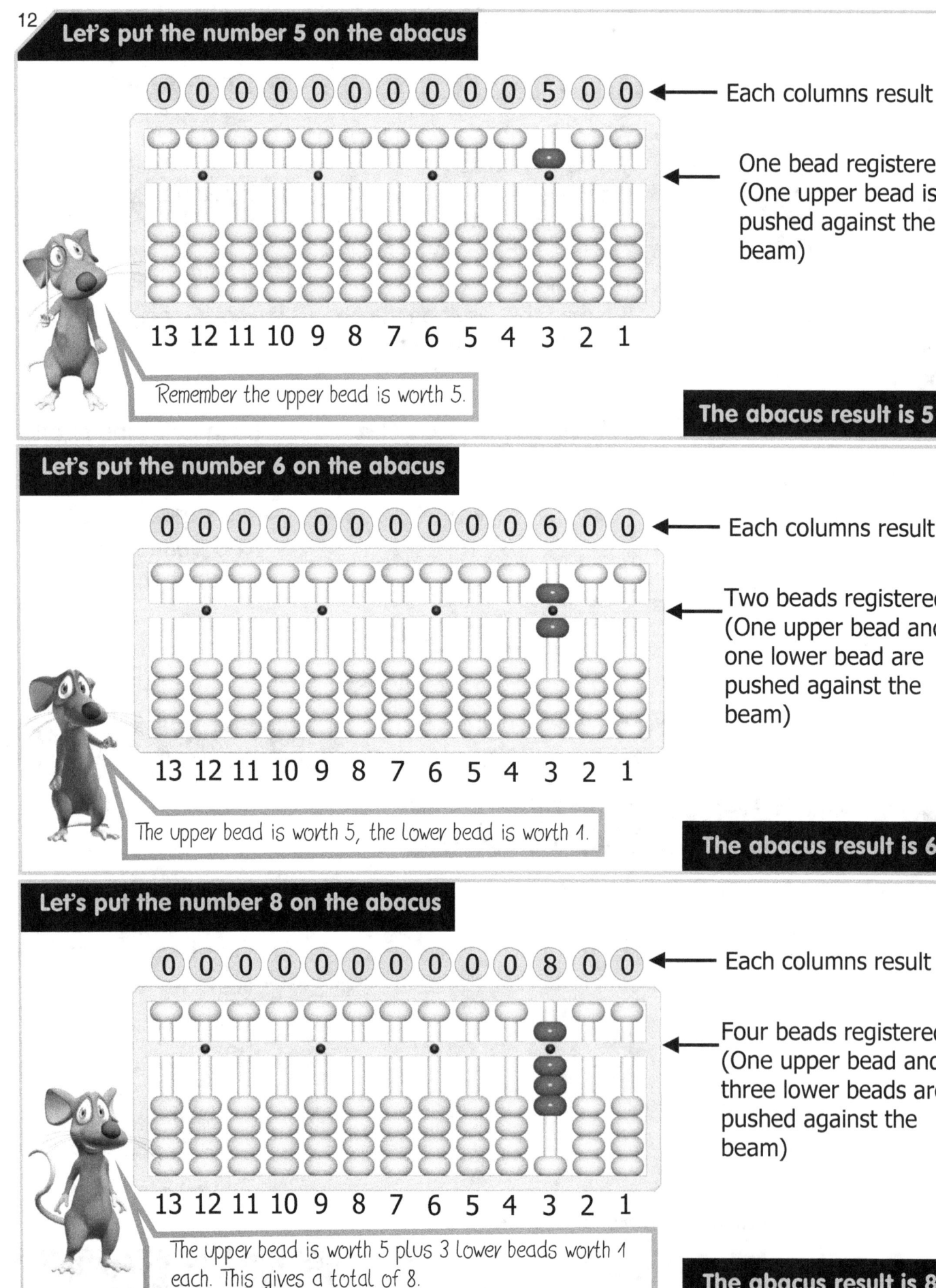

Let's put the number 5 on the abacus

0 0 0 0 0 0 0 0 0 0 5 0 0 ← Each columns result

← One bead registered (One upper bead is pushed against the beam)

13 12 11 10 9 8 7 6 5 4 3 2 1

Remember the upper bead is worth 5.

The abacus result is 5

Let's put the number 6 on the abacus

0 0 0 0 0 0 0 0 0 0 6 0 0 ← Each columns result

← Two beads registered (One upper bead and one lower bead are pushed against the beam)

13 12 11 10 9 8 7 6 5 4 3 2 1

The upper bead is worth 5, the lower bead is worth 1.

The abacus result is 6

Let's put the number 8 on the abacus

0 0 0 0 0 0 0 0 0 0 8 0 0 ← Each columns result

← Four beads registered (One upper bead and three lower beads are pushed against the beam)

13 12 11 10 9 8 7 6 5 4 3 2 1

The upper bead is worth 5 plus 3 lower beads worth 1 each. This gives a total of 8.

The abacus result is 8

Here are the single digit numbers on the abacus

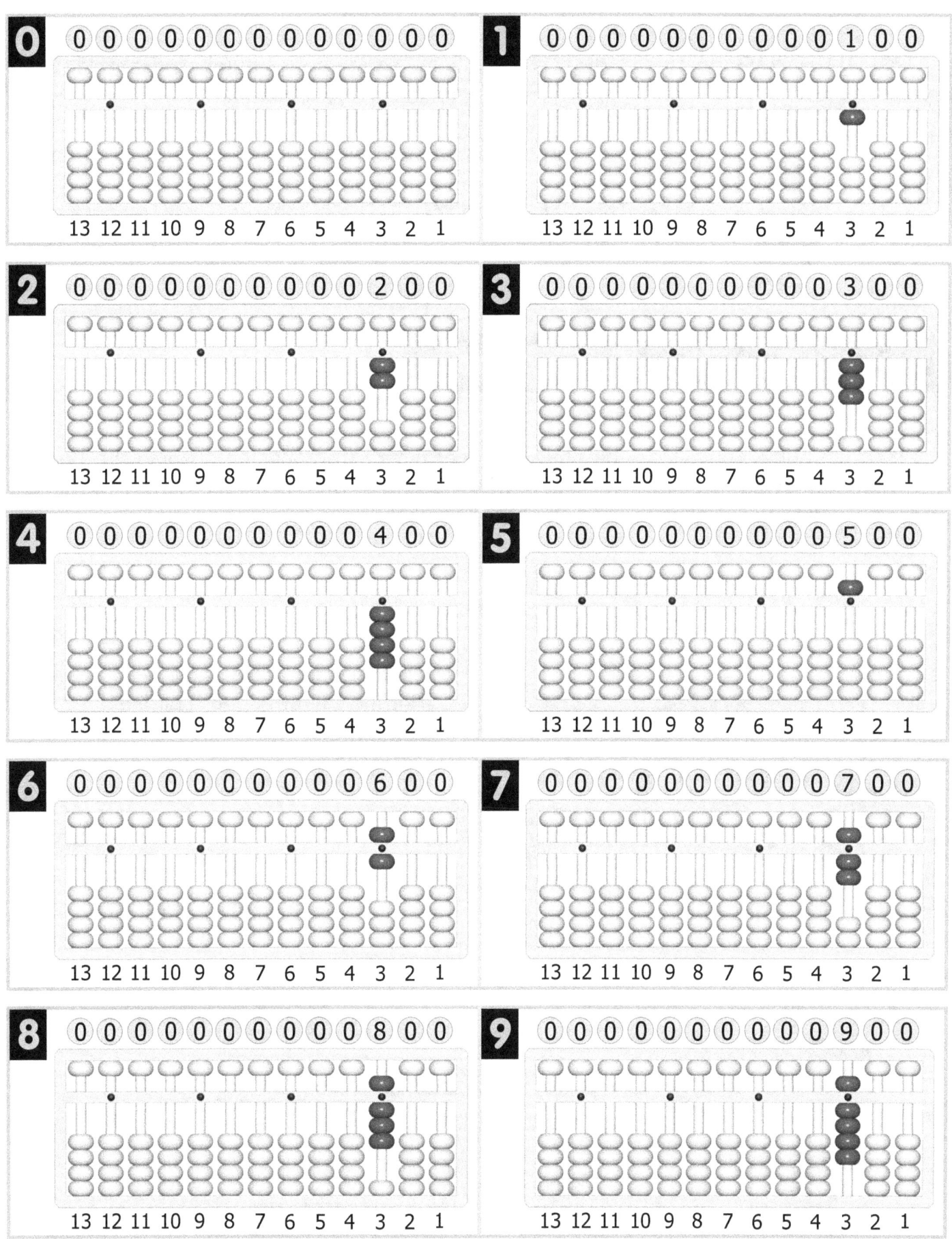

HOW TO PUT A NUMBER ON THE ABACUS WITH MORE THAN ONE DIGIT

Important!

- A **digit** is a symbol used to show a number. Example, **6** is one digit and is made up of one number.
- A digit is any number from **0** to **9**.
- The number **78** has two digits, **7** and **8**. There are two digits that make up the number 78.
- There are two places in the number 78, the ones place holds the number 8 and the tens place holds the number 7.

Let's put the number 15 on the abacus

When we put a number on the abacus or '**register**' a number by pushing the beads towards the beam, we start with the leftmost digit first (in this example the 1 of the 15) then move to the right to register the other numbers.

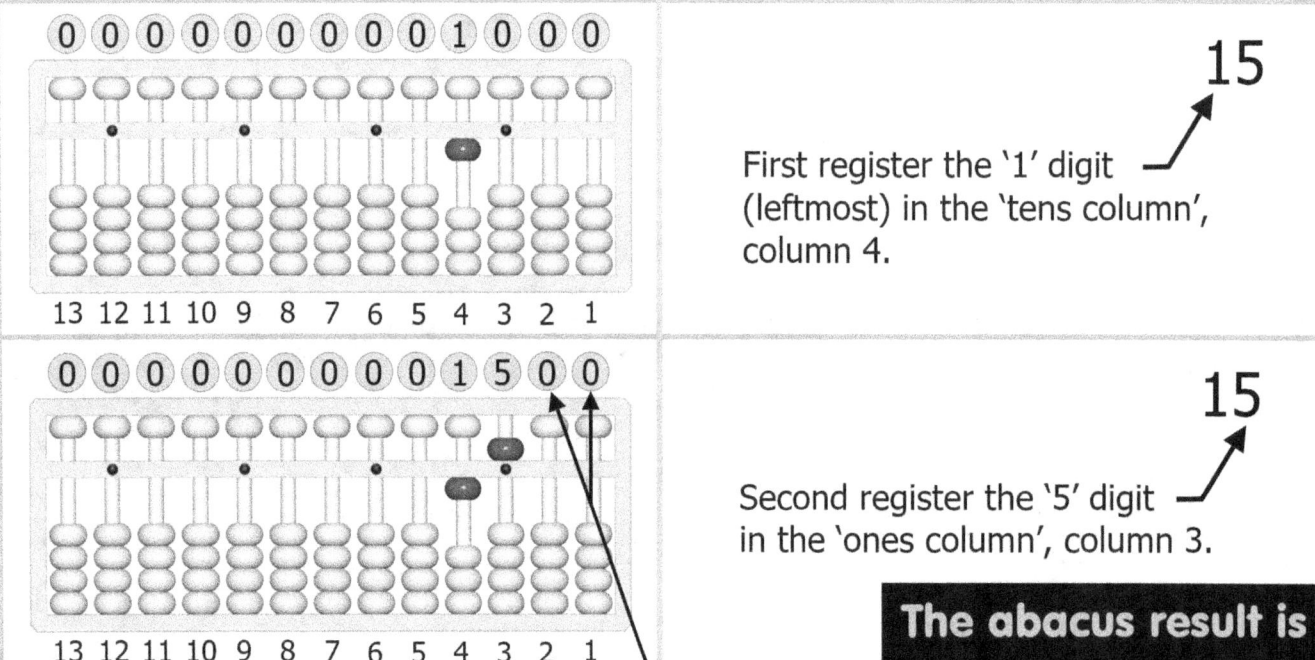

15

First register the '1' digit (leftmost) in the 'tens column', column 4.

15

Second register the '5' digit in the 'ones column', column 3.

The abacus result is 15

The two zeros after the number 15 are for decimal numbers, tenths (example the digit 3 in 0.3) for column 2 and hundredths in column 1 (example the digit 6 in 0.46).
We can **ignore** them as we are using whole numbers (also called counting numbers 1,2,3,4....).

Here are some two digit numbers on the abacus

45

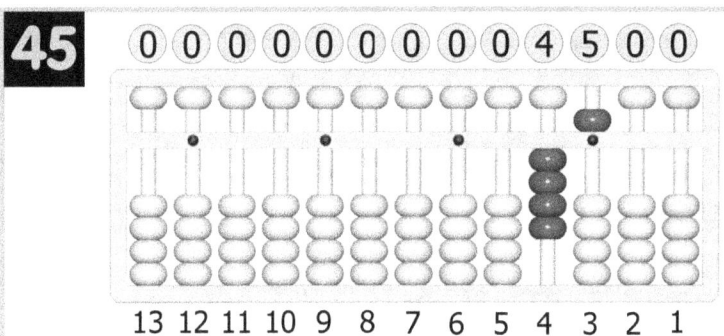

0 0 0 0 0 0 0 0 0 4 5 0 0

13 12 11 10 9 8 7 6 5 4 3 2 1

- We will put 45 on the abacus
- 45 has 2 digits, so use 2 columns
- Column 4, register 4 lower beads
  (this is for the 4 of the 45)
- Column 3, register 1 upper bead
  (this is for the 5 of the 45)

The abacus result is 45

14

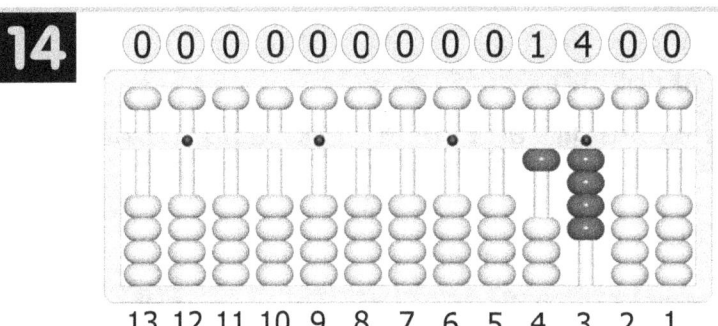

0 0 0 0 0 0 0 0 0 1 4 0 0

13 12 11 10 9 8 7 6 5 4 3 2 1

- We will put 14 on the abacus
- 14 has 2 digits, so use 2 columns
- Column 4, register 1 lower bead
  (this is for the 1 of the 14)
- Column 3, register 4 lower beads
  (this is for the 4 of the 14)

The abacus result is 14

95

0 0 0 0 0 0 0 0 0 9 5 0 0

13 12 11 10 9 8 7 6 5 4 3 2 1

- We will put 95 on the abacus
- 95 has 2 digits, so use 2 columns
- Column 4, register 1 upper bead and 4
  lower beads (this is for the 9 of the 95)
  Total on this column is 5+4=9
- Column 3, register 1 upper bead
  (this is for the 5 of the 95)

The abacus result is 95

56

0 0 0 0 0 0 0 0 0 5 6 0 0

13 12 11 10 9 8 7 6 5 4 3 2 1

- We will put 56 on the abacus
- 56 has 2 digits, so use 2 columns
- Column 4, register 1 upper bead
  (this is for the 5 of the 56)
- Column 3, register 1 upper bead and 1
  lower bead (this is for the 6 of the 56)
  Total on this column is 5+1=6

The abacus result is 56

87

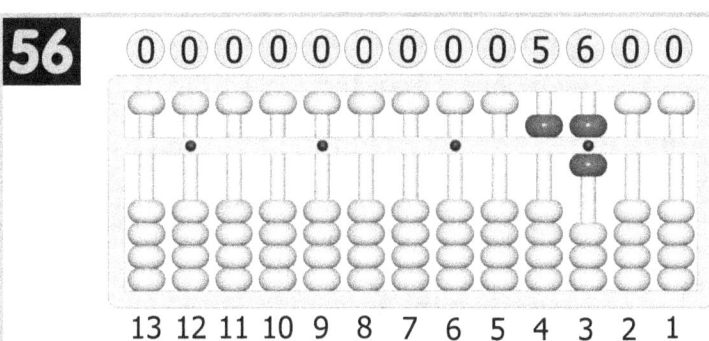

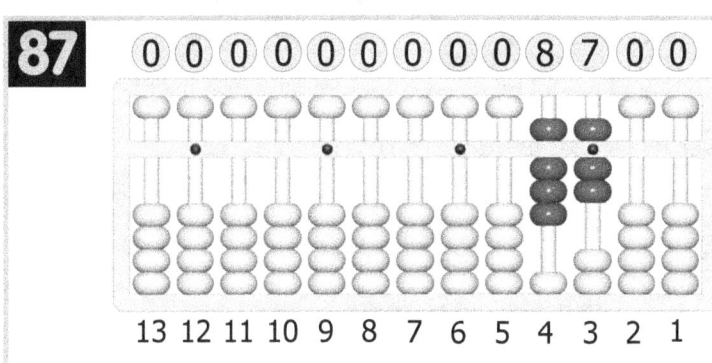

0 0 0 0 0 0 0 0 8 7 0 0

13 12 11 10 9 8 7 6 5 4 3 2 1

- We will put 87 on the abacus
- 87 has 2 digits, so use 2 columns
- Column 4, register 1 upper bead and 3
  lower beads (this is for the 8 of the 87).
  Total on this column is 5+3=8
- Column 3, register 1 upper bead and 2
  lower beads (this is for the 7 of the 87).
  Total on this column is 5+2=7

The abacus result is 87

How to register multi-digit numbers on the abacus

A multi-digit number is any number that has more than one digit.

We have already looked at some multi-digit numbers on the previous page (two digit numbers).

Now we will look at larger numbers. Let's start with a 5 digit number.

23456

First register the '2' digit (leftmost) in the 7th column.

Why the 7th column? Because the number has 5 digits and we are not using the first 2 columns.

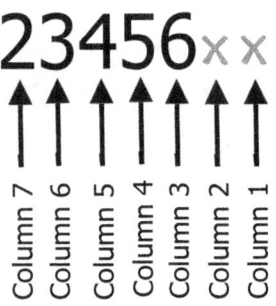

23456 x x

Column 7
Column 6
Column 5
Column 4
Column 3
Column 2
Column 1

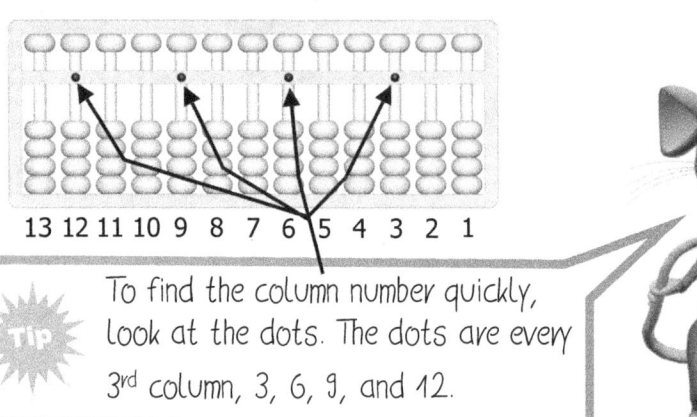

13 12 11 10 9 8 7 6 5 4 3 2 1

Tip

To find the column number quickly, look at the dots. The dots are every 3rd column, 3, 6, 9, and 12.

23456 Column 6 dot

0 0 0 0 0 0 2 3 4 5 6 0 0

13 12 11 10 9 8 7 6 5 4 3 2 1

- We will put 23456 on the abacus
- 23456 has 5 digits, so use 5 columns (start on column 7)

- Column 7, register 2 lower beads
- Column 6, register 3 lower beads
- Column 5, register 4 lower beads
- Column 4, register 1 upper bead
- Column 3, register 1 upper bead and 1 lower bead
  (total on this column is 5+1=6)

The abacus result is 23456

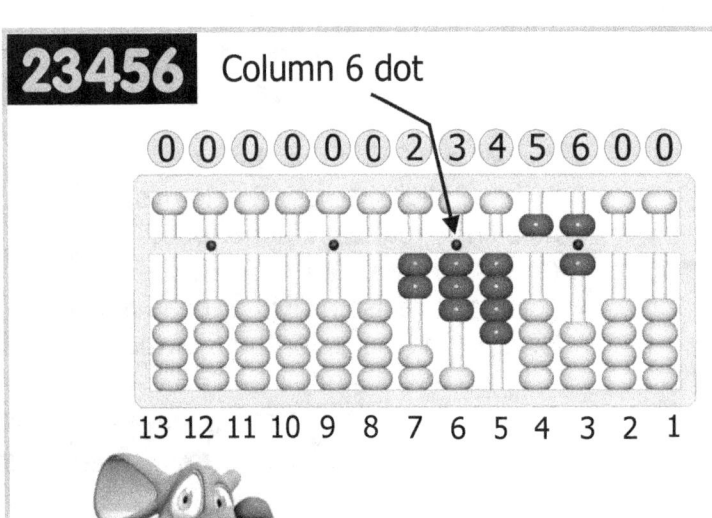

Things to remember before we move on:
- Don't use columns 1 and 2 (keep those for decimal numbers)
- The total digits of the number plus 2 = the column where we start to register our number
- The dots help us find the column number

Here are some multi-digit numbers on the abacus

556677

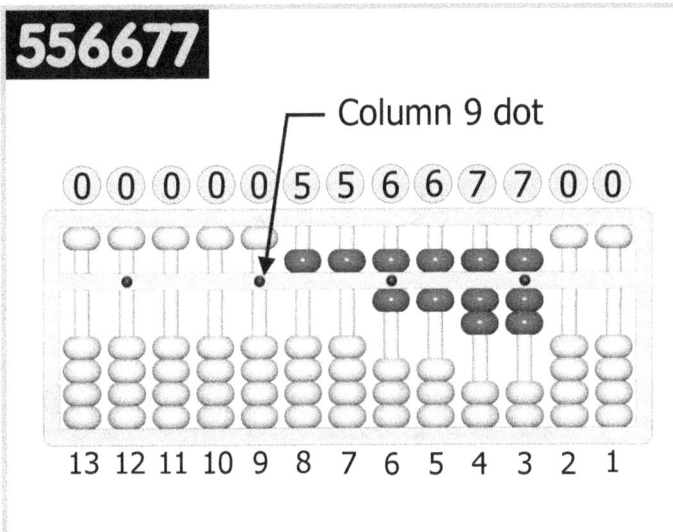

Column 9 dot

0 0 0 0 0 5 5 6 6 7 7 0 0

13 12 11 10 9 8 7 6 5 4 3 2 1

- We will put 556677 on the abacus
- 556677 has 6 digits, so use 6 columns (start on column 8)
- Column 8, register 1 higher bead
- Column 7, register 1 higher bead
- Column 6, register 1 upper bead and 1 lower bead
- Column 5, register 1 upper bead and 1 lower bead
- Column 4, register 1 upper bead and 2 lower beads
- Column 3, register 1 upper bead and 2 lower beads

The abacus result is 556677

316

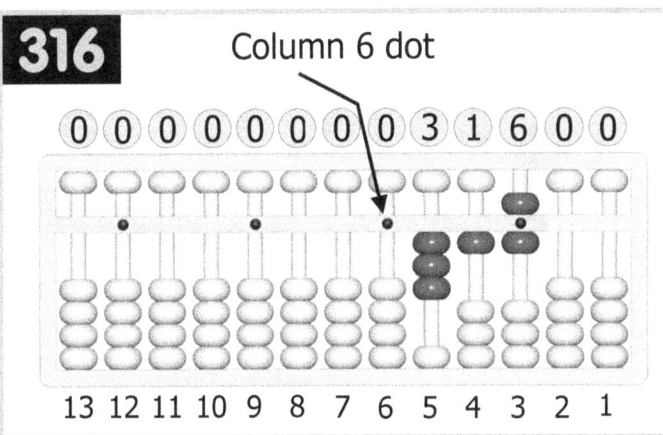

Column 6 dot

0 0 0 0 0 0 0 3 1 6 0 0

13 12 11 10 9 8 7 6 5 4 3 2 1

- We will put 316 on the abacus
- 316 has 3 digits, so use 3 columns (start on column 5)
- Column 5, register 3 lower beads
- Column 4, register 1 lower bead
- Column 3, register 1 upper bead and 1 lower bead

The abacus result is 316

920030598

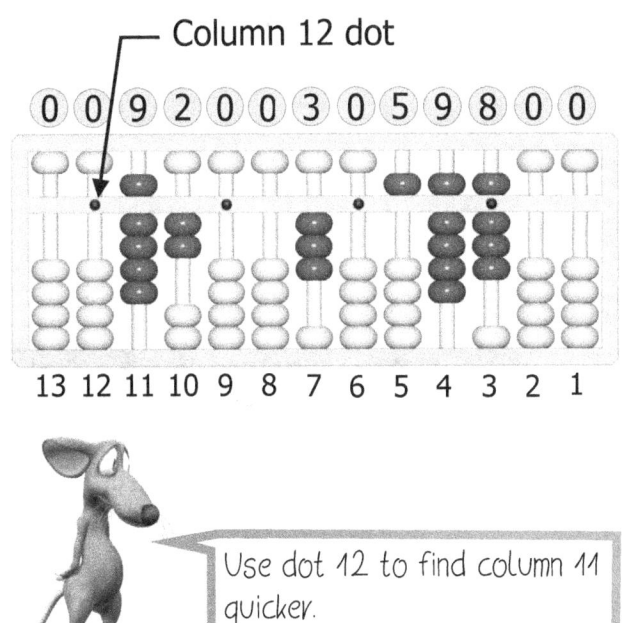

Column 12 dot

0 0 9 2 0 0 3 0 5 9 8 0 0

13 12 11 10 9 8 7 6 5 4 3 2 1

- We will put 920030598 on the abacus
- 920030598 has 9 digits, so use 9 columns (start on column 11)
- Column 11, register 1 upper bead and 4 lower beads
- Column 10, register 2 lower beads
- Column 9, do nothing
- Column 8, do nothing
- Column 7, register 3 lower beads
- Column 6, do nothing
- Column 5, register 1 upper bead
- Column 4, register 1 upper bead and 4 lower beads
- Column 3, register 1 upper bead and 3 lower beads

Use dot 12 to find column 11 quicker.

Columns 9, 8 & 6 were easy!

The abacus result is 920030598

TEST 1 - What numbers are shown on the abacus?

Can you read the number on each abacus?
If you can, write the number in the box next to it.

Answers are on page 20.

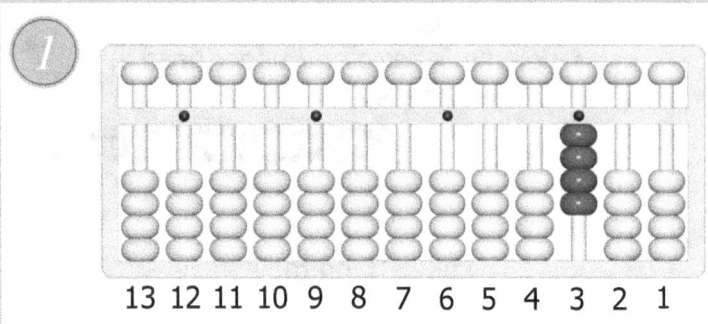

What number is on the abacus?

13 12 11 10 9 8 7 6 5 4 3 2 1

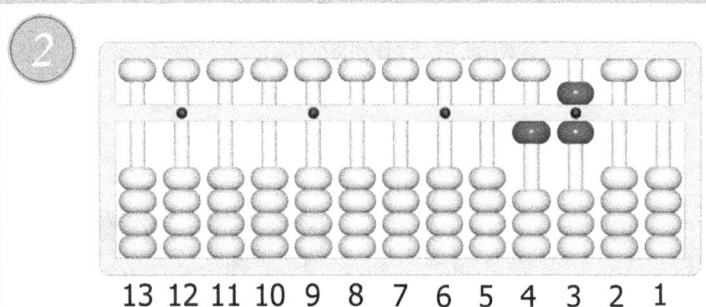

What number is on the abacus?

13 12 11 10 9 8 7 6 5 4 3 2 1

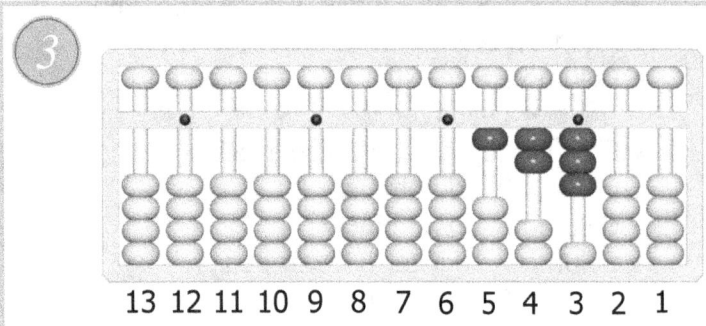

What number is on the abacus?

13 12 11 10 9 8 7 6 5 4 3 2 1

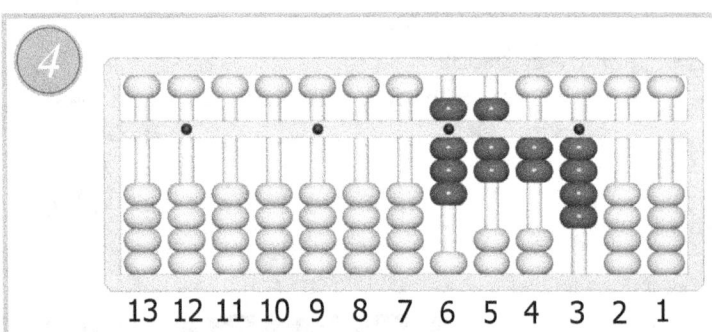

What number is on the abacus?

13 12 11 10 9 8 7 6 5 4 3 2 1

TEST 1 - What numbers are shown on the abacus?

Answers are on page 20.

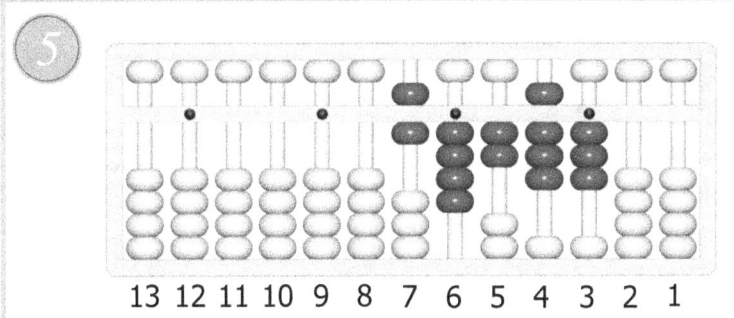

What number is on the abacus?

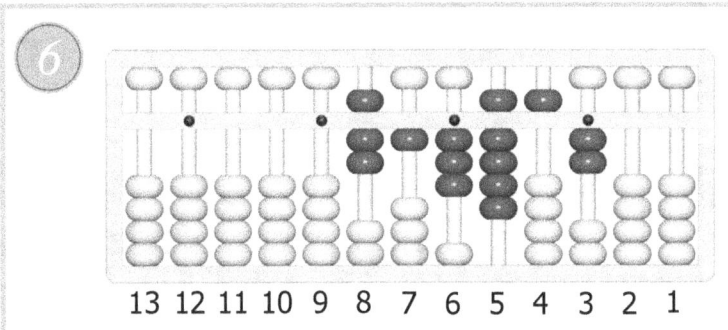

What number is on the abacus?

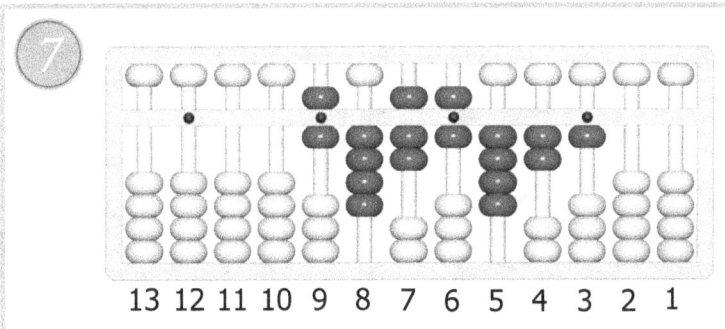

What number is on the abacus?

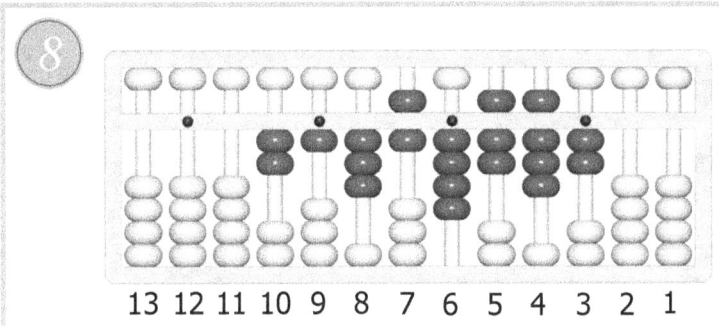

What number is on the abacus?

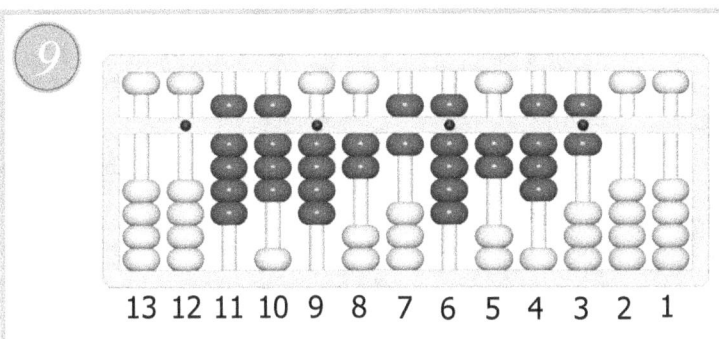

What number is on the abacus?

Answers to test 1 (on pages 18 & 19)

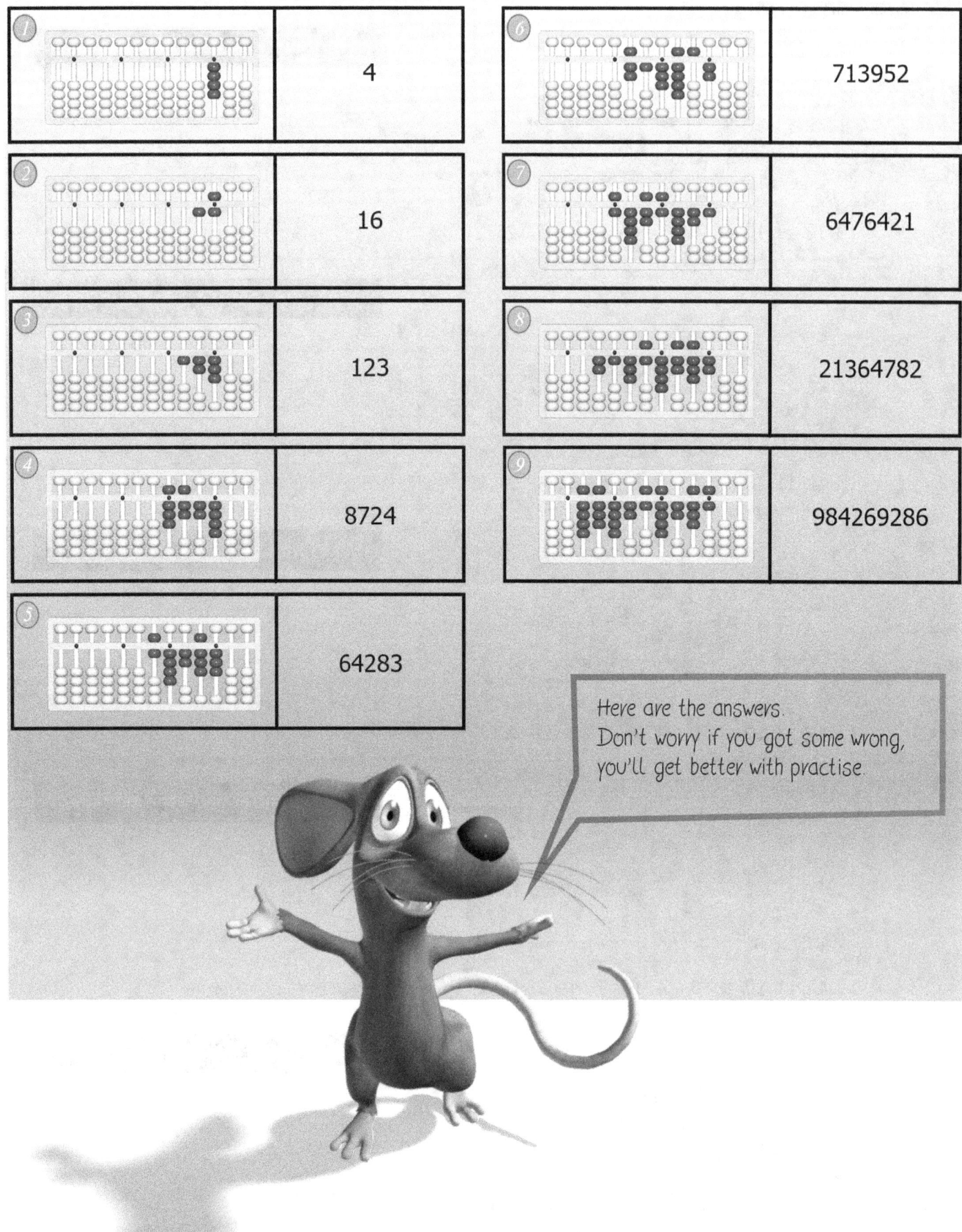

1	4
2	16
3	123
4	8724
5	64283
6	713952
7	6476421
8	21364782
9	984269286

Here are the answers.
Don't worry if you got some wrong,
you'll get better with practise.

MOVING THE BEADS
What fingers do you use to move the beads?

The pictures on pages 21 to 25 will show you what fingers you use to register and unregister the beads.

We will use the thumb, index and middle fingers. Some people like to use only the thumb and index finger, but we will use the middle finger also.

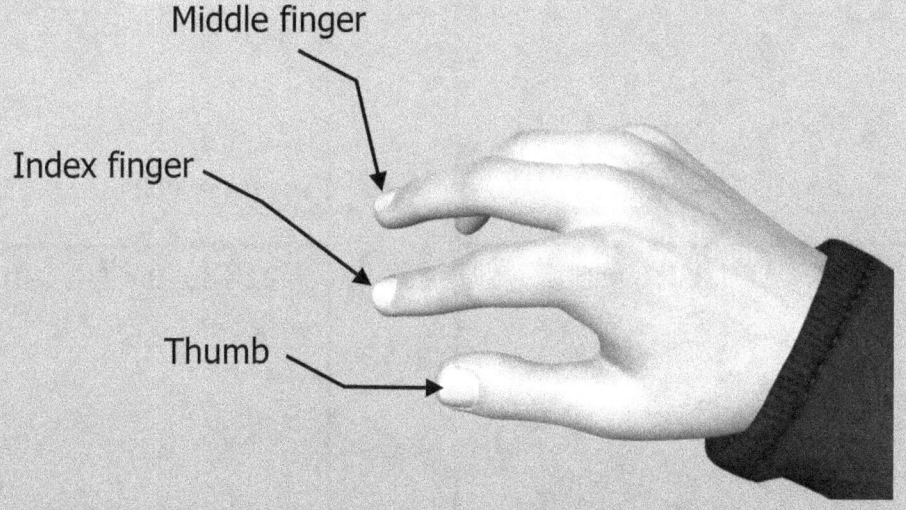

Middle finger

Index finger

Thumb

- **Thumb**
  Used to register the **LOWER** beads (to push them towards the beam).

- **Index finger**
  Used to unregister the **LOWER** beads (to move them away from the beam).

- **Middle finger**
  Used to register and unregister the **UPPER** beads (to move them to and away from the beam).

- **Bead order**
  When registering and unregistering a number, always start with the highest value column first, then work towards the lowest value column.
  For example, when registering the number 123 start by registering the 1 (100) then the 2 (20) and finally the 3 (3).

On the next 4 pages there are pictures which show the finger movements for registering and unregistering.

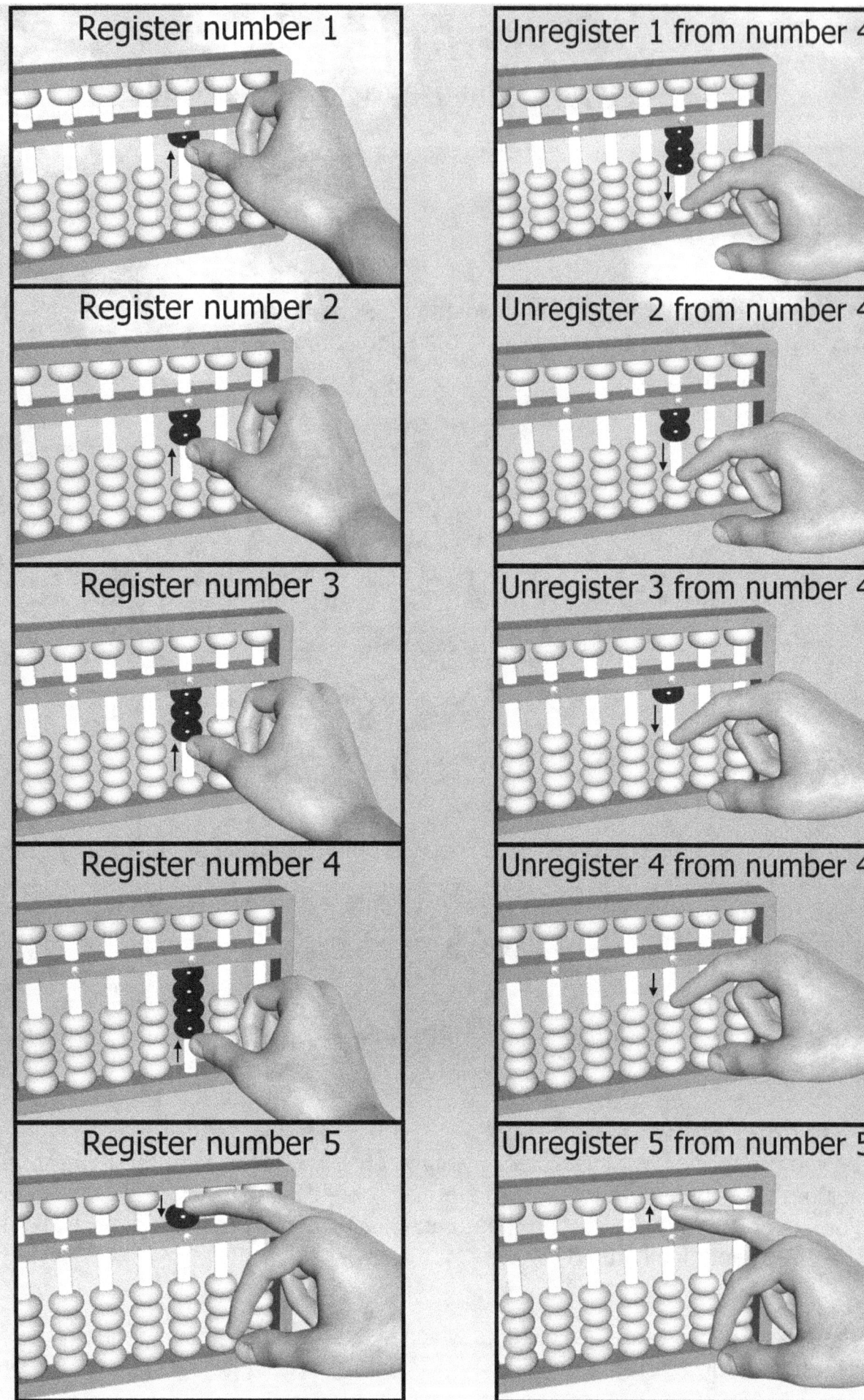

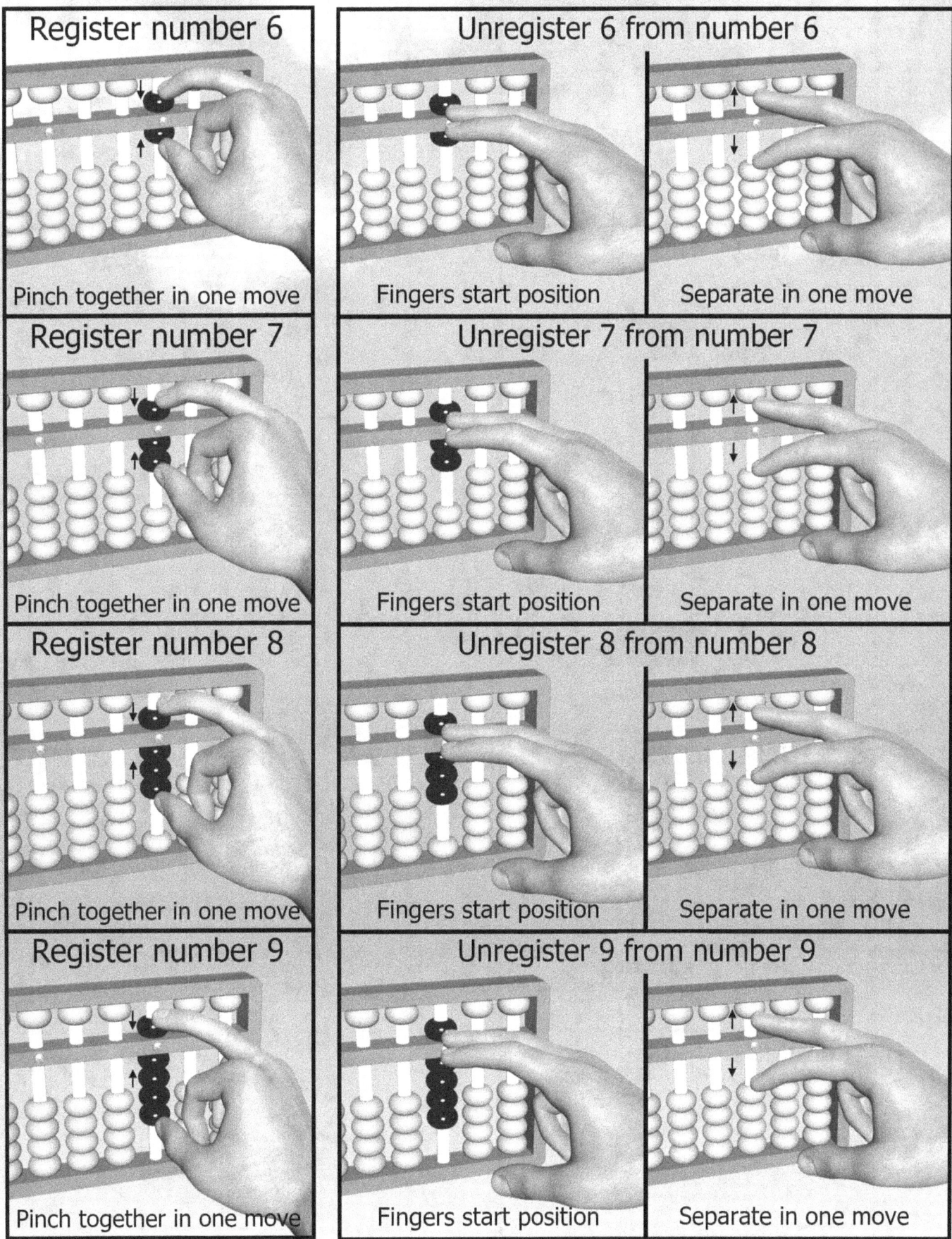

23

Register number 6
Pinch together in one move

Unregister 6 from number 6
Fingers start position
Separate in one move

Register number 7
Pinch together in one move

Unregister 7 from number 7
Fingers start position
Separate in one move

Register number 8
Pinch together in one move

Unregister 8 from number 8
Fingers start position
Separate in one move

Register number 9
Pinch together in one move

Unregister 9 from number 9
Fingers start position
Separate in one move

24

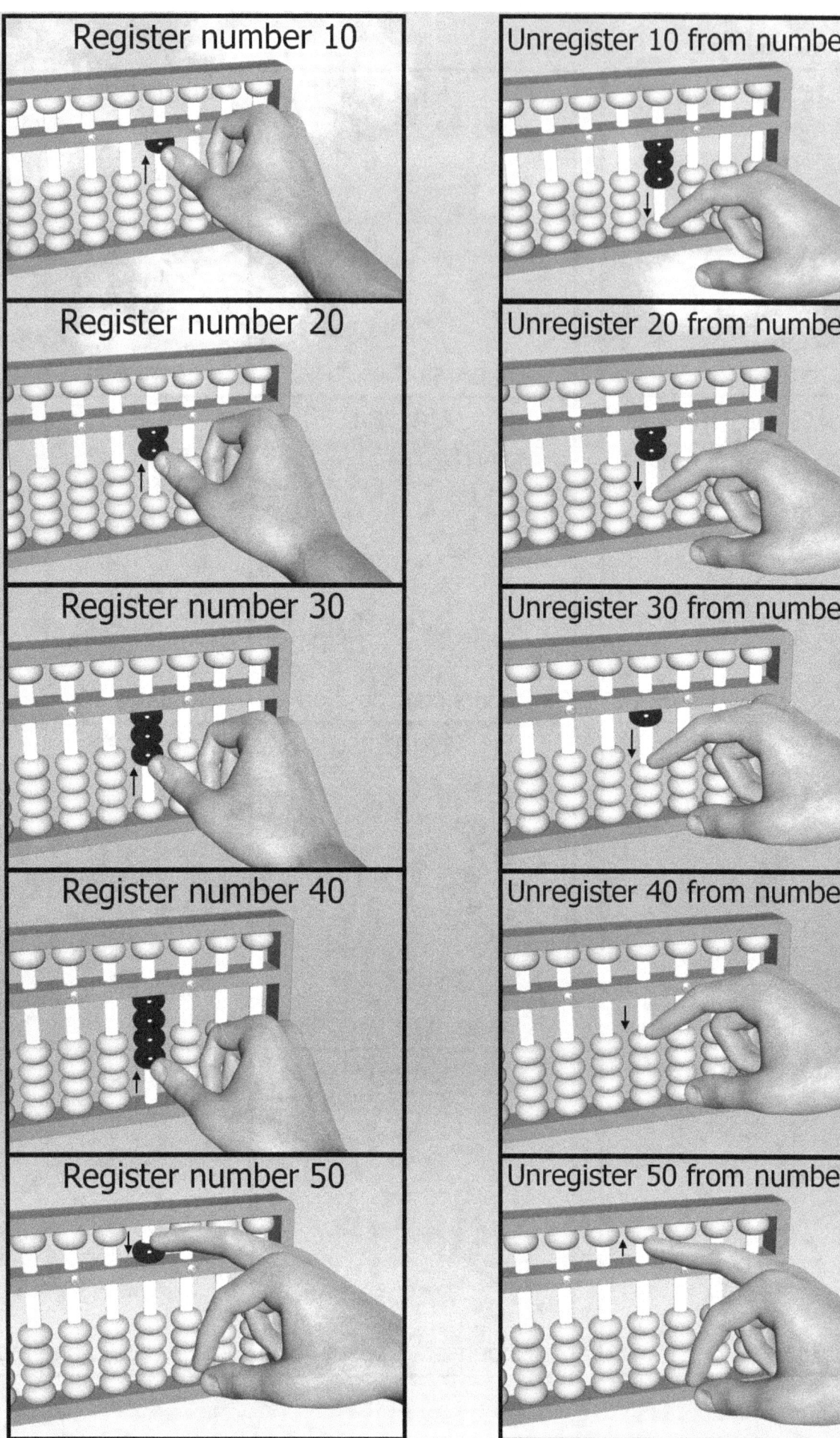

Register number 10	Unregister 10 from number 40
Register number 20	Unregister 20 from number 40
Register number 30	Unregister 30 from number 40
Register number 40	Unregister 40 from number 40
Register number 50	Unregister 50 from number 50

25

Register the number 12

First register 10

Second register 2

Unregister 12 from the number 12

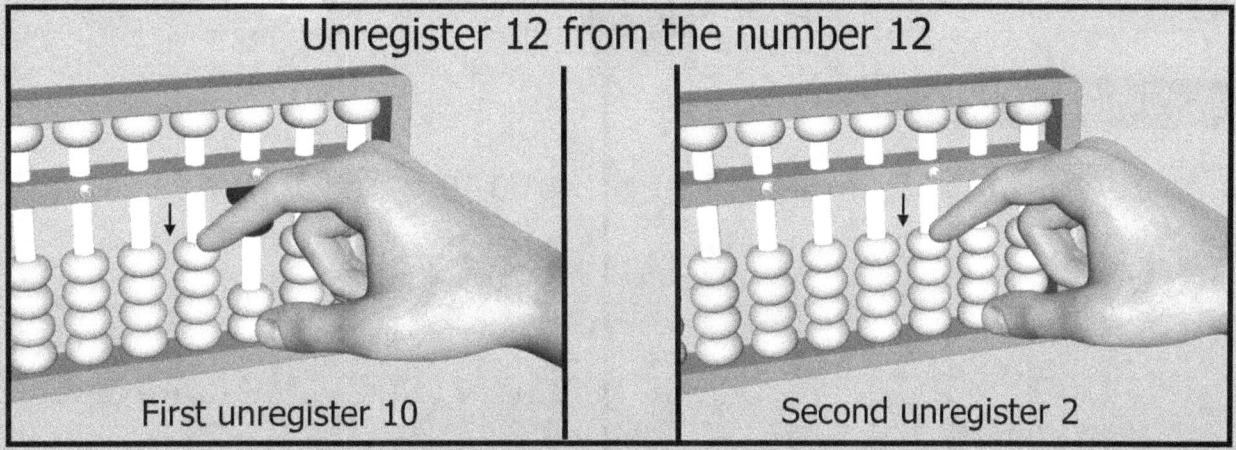

First unregister 10

Second unregister 2

Register the number 39

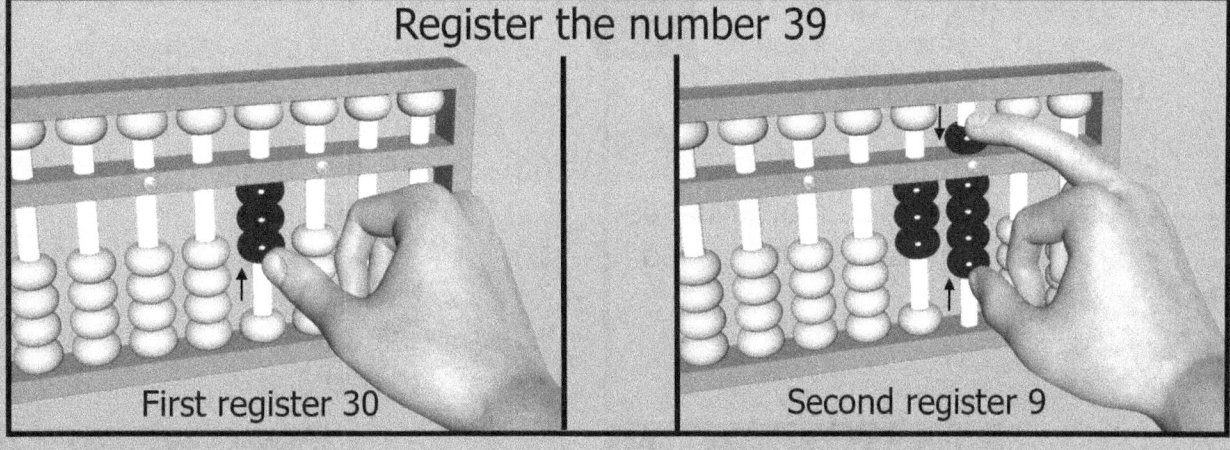

First register 30

Second register 9

Unregister 12 from the number 39 = 27

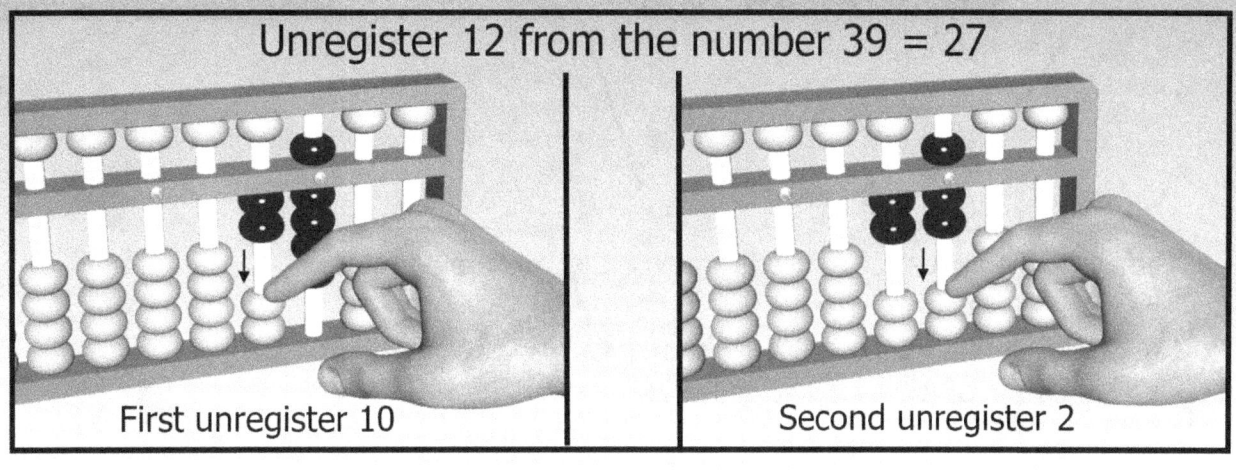

First unregister 10

Second unregister 2

ADDITION

> Addition is adding numbers to get the sum of those numbers

Addition - things to remember:
- Register your numbers from left to right, for example: for number 231 register the 2 first, 3 second and 1 last.
- Each digit must be registered in the correct column, for example with 231 the 2 is for column 5 (hundredths column), the 3 for column 4 (tens column) and the 1 for column 3 (ones column).

Example: 231 + 213

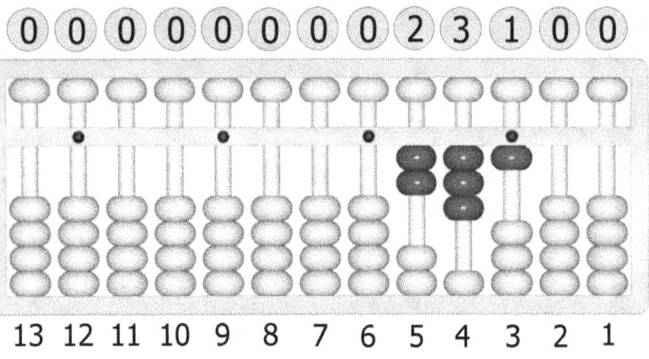

0 0 0 0 0 0 0 2 3 1 0 0

13 12 11 10 9 8 7 6 5 4 3 2 1

We will **register 231**

200 • Column 5, register 2 lower beads
30 • Column 4, register 3 lower beads
1 • Column 3, register 1 lower bead

The abacus reads 231

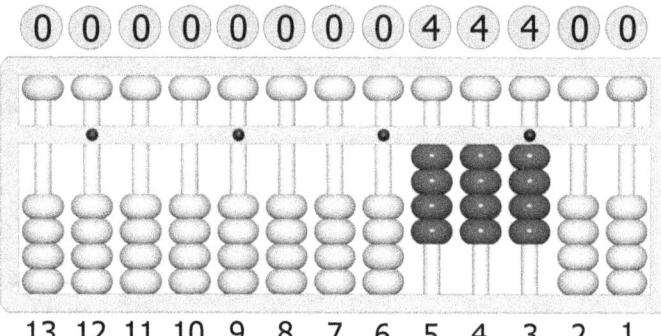

0 0 0 0 0 0 0 0 4 4 4 0 0

13 12 11 10 9 8 7 6 5 4 3 2 1

We will now **add 213** to 231

+200 • Column 5, register 2 lower beads to add 200
+10 • Column 4, register 1 lower bead to add 10
+3 • Column 3, register 3 lower beads to add 3

The abacus result is 444

> These columns are useful to see the amount that you are adding.
> For example:
>   30 means that you have just registered 30
>   +200 means that you have just added 200

More addition examples

Example: 4 + 5

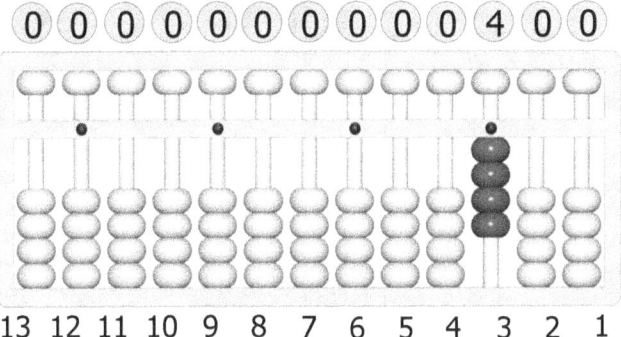

0 0 0 0 0 0 0 0 0 4 0 0

13 12 11 10 9 8 7 6 5 4 3 2 1

We will **register 4**

4
- Column 3, register 4 lower beads

The abacus reads 4

---

0 0 0 0 0 0 0 0 0 9 0 0

13 12 11 10 9 8 7 6 5 4 3 2 1

We will now **add 5** to 4

+5
- Column 3, register 1 upper bead to add 5

The abacus result is 9

Remember!
4 means that you have just registered 4
+5 means that you have just added 5

Example: 56 + 23

0 0 0 0 0 0 0 0 0 5 6 0 0

13 12 11 10 9 8 7 6 5 4 3 2 1

We will **register 56**

50
- Column 4, register 1 upper bead (this is the 5 of the 56)

6
- Column 3, register 1 upper bead and 1 lower bead (this is the 6 of the 56)

The abacus reads 56

---

0 0 0 0 0 0 0 0 0 7 9 0 0

13 12 11 10 9 8 7 6 5 4 3 2 1

We will now **add 23** to 56

+20
- Column 4, register 2 lower beads to add 20

+3
- Column 3, register 3 lower beads to add 3

The abacus result is 79

More addition examples

Example: 527 + 462

0 0 0 0 0 0 0 0 5 2 7 0 0

13 12 11 10 9 8 7 6 5 4 3 2 1

We will **register 527**

500 • Column 5, register 1 upper bead (this is the 5 of the 527)

20 • Column 4, register 2 lower beads (this is the 2 of the 527)

7 • Column 3, register 1 upper bead and 2 lower beads (this is the 7 of the 527)

The abacus reads 527

0 0 0 0 0 0 0 0 9 8 9 0 0

13 12 11 10 9 8 7 6 5 4 3 2 1

We will now **add 462** to 527

+400 • Column 5, register 4 lower beads to add 400

+60 • Column 4, register 1 upper bead and 1 lower bead to add 60

+2 • Column 3, register 2 lower beads to add 2

The abacus result is 989

Example: 2112 + 32

0 0 0 0 0 0 0 2 1 1 2 0 0

13 12 11 10 9 8 7 6 5 4 3 2 1

We will **register 2112**

2000 • Column 6, register 2 lower beads

100 • Column 5, register 1 lower bead

10 • Column 4, register 1 lower bead

2 • Column 3, register 2 lower beads

The abacus reads 2112

0 0 0 0 0 0 0 2 1 4 4 0 0

13 12 11 10 9 8 7 6 5 4 3 2 1

We will **add 32**

+30 • Column 4, register 3 lower beads

+2 • Column 3, register 2 lower beads

The abacus result is 2144

More addition examples

Example: 65260 + 4539

0	0	0	0	0	0	6	5	2	6	0	0	0

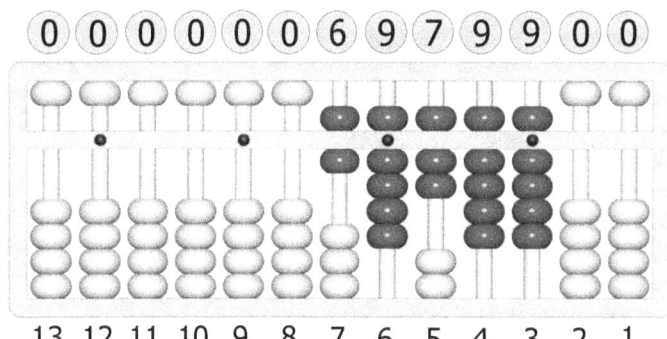

13 12 11 10 9 8 7 6 5 4 3 2 1

We will **register 65260**

60000	• Column 7, register 1 upper bead and 1 lower bead
5000	• Column 6, register 1 upper bead
200	• Column 5, register 2 lower beads
60	• Column 4, register 1 upper bead and 1 lower bead
	• Column 3, do nothing

The abacus reads 65260

0	0	0	0	0	0	6	9	7	9	9	0	0

13 12 11 10 9 8 7 6 5 4 3 2 1

We will now **add 4539** to 65260

+4000	• Column 6, register 4 lower beads to add 4000
+500	• Column 5, register 1 upper bead to add 500
+30	• Column 4, register 3 lower beads to add 30
+9	• Column 3, register 1 upper bead and 4 lower beads to add 9

The abacus result is 69799

Example: 123456789 + 121521210

0	0	1	2	3	4	5	6	7	8	9	0	0

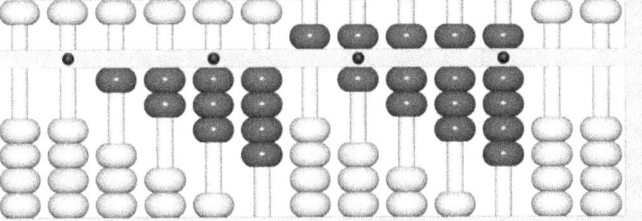

13 12 11 10 9 8 7 6 5 4 3 2 1

We will **register 123456789**

100000000	• Column 11, register 1 lower bead
20000000	• Column 10, register 2 lower beads
3000000	• Column 9, register 3 lower beads
400000	• Column 8, register 4 lower beads
50000	• Column 7, register 1 upper bead
6000	• Column 6, register 1 upper and 1 lower bead
700	• Column 5, register 1 upper and 2 lower beads
80	• Column 4, register 1 upper and 3 lower beads
9	• Column 3, register 1 upper and 4 lower beads

The abacus reads 123456789

0	0	2	4	4	9	7	7	9	9	9	0	0

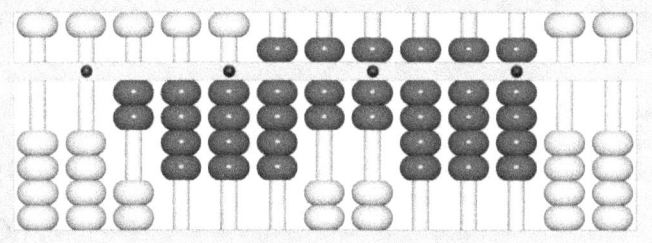

13 12 11 10 9 8 7 6 5 4 3 2 1

We will **add 121521210**

+100000000	• Column 11, register 1 lower bead
+20000000	• Column 10, register 2 lower beads
+1000000	• Column 9, register 1 lower bead
+500000	• Column 8, register 1 upper bead
+20000	• Column 7, register 2 lower beads
+1000	• Column 6, register 1 lower bead
+200	• Column 5, register 2 lower beads
+10	• Column 4, register 1 lower bead
	• Column 3, do nothing

The abacus result is 244977999

More addition examples

Example: 523602092 + 425160905

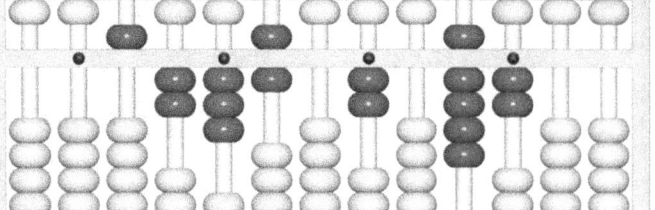

13	12	11	10	9	8	7	6	5	4	3	2	1

We will register 523602092

500000000	• Column 11, register 1 upper bead
20000000	• Column 10, register 2 lower beads
3000000	• Column 9, register 3 lower beads
600000	• Column 8, register 1 upper and 1 lower bead
	• Column 7, do nothing
2000	• Column 6, register 2 lower beads
	• Column 5, do nothing
90	• Column 4, register 1 upper and 4 lower beads
2	• Column 3, register 2 lower beads

The abacus reads 523602092

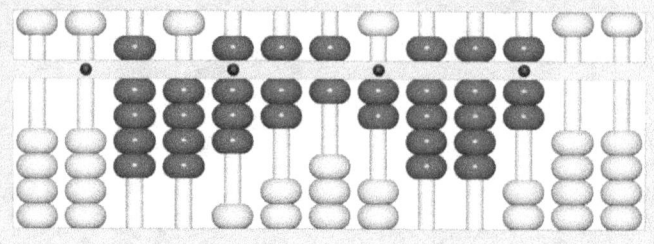

13	12	11	10	9	8	7	6	5	4	3	2	1

We will add 425160905

+400000000	• Column 11, register 4 lower beads
+20000000	• Column 10, register 2 lower beads
+5000000	• Column 9, register 1 upper bead
+100000	• Column 8, register 1 lower bead
+60000	• Column 7, register 1 upper and 1 lower bead
	• Column 6, do nothing
+900	• Column 5, register 1 upper and 4 lower beads
	• Column 4, do nothing
+5	• Column 3, register 1 upper bead

The abacus result is 948762997

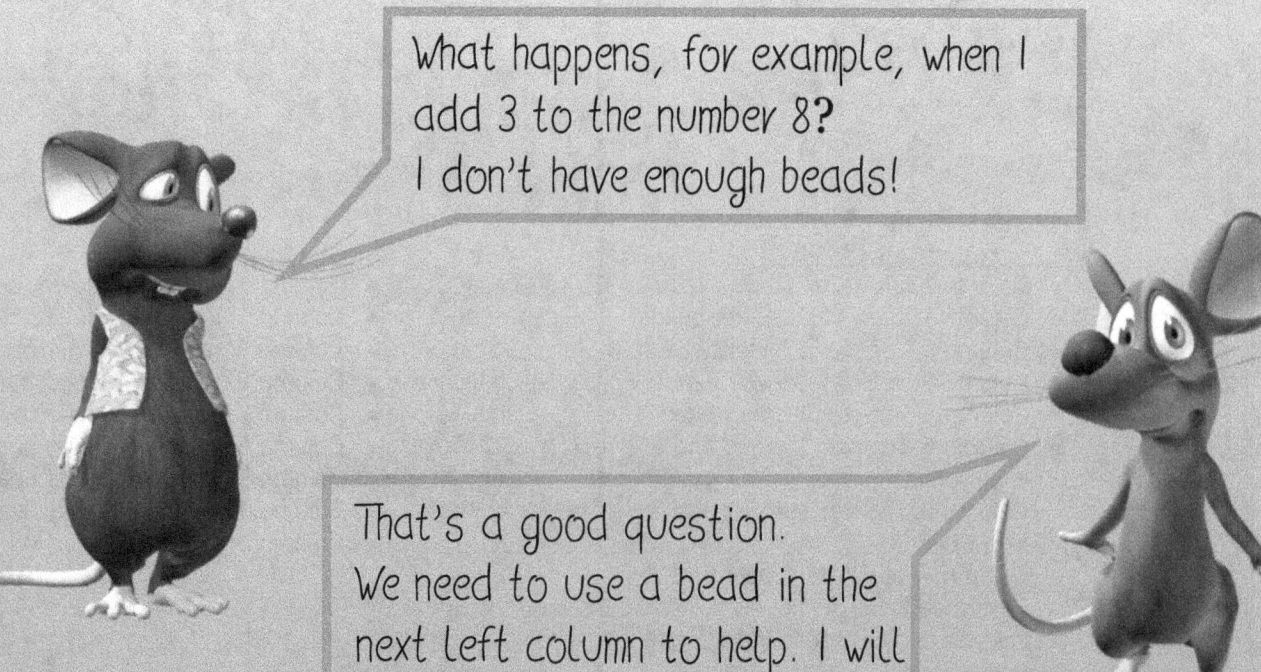

What happens, for example, when I add 3 to the number 8? I don't have enough beads!

That's a good question. We need to use a bead in the next left column to help. I will show you on the next page.

When there are not enough beads in the column for the addition

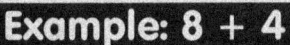

> When you don't have enough beads, move to the next LEFT column to help

For example, when you try to add 4 to the already registered number 8, you don't have enough beads in the column to do it. You can only register a maximum of 9 in each column (4 lower beads and 1 upper bead, 4+5=9).
When this happens, we need to use the
'**Not enough beads list**'.

1=10-9
2=10-8
3=10-7
4=10-6
5=10-5
6=10-4
7=10-3
8=10-2
9=10-1

How to use the 'Not enough beads list'

Let's say we need to add 3 to a column but we don't have enough beads.

Look at the list, **3=10-7**

10 is the number to **register,** in the next **LEFT** column (1 lower bead).

7 is the number to **unregister** in our column.

Example: 8 + 4

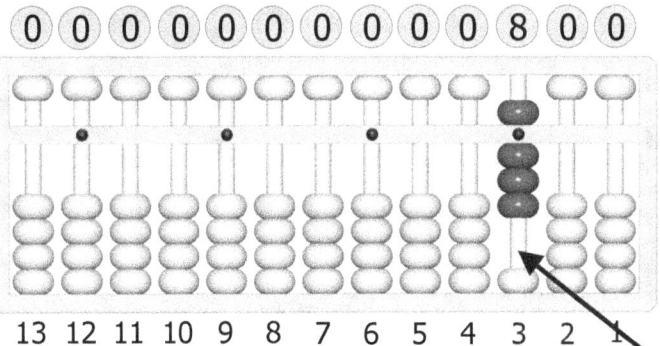

0 0 0 0 0 0 0 0 0 0 8 0 0

13 12 11 10 9 8 7 6 5 4 3 2 1

We will **register 8**

8 • Column 3, register 1 upper bead and 3 lower beads

> Unregister means move away from the beam (subtract)!

The abacus reads 8

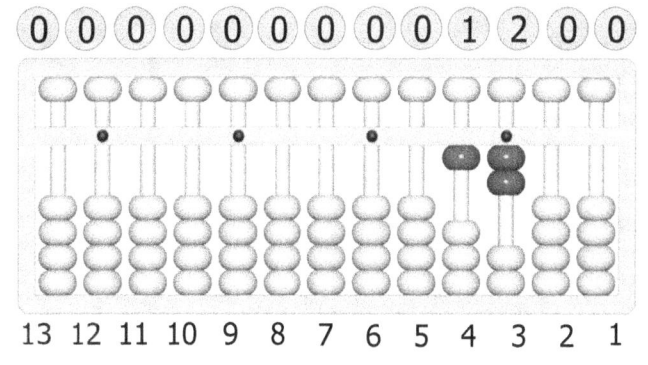

0 0 0 0 0 0 0 0 0 0 1 2 0 0

13 12 11 10 9 8 7 6 5 4 3 2 1

We will now **add 4**

• There are not enough beads in column 3 to add 4
• First we must think that **4=10-6** see the '**Not enough beads list**'
-6 • Column 3, **unregister** 1 upper and 1 lower bead to subtract 6 (4=10-**6**)
+10 • Column 4, register 1 lower bead to add 10 (4=**10**-6)

The abacus result is 12

More addition examples (when we don't have enough beads)

Example: 9 + 5

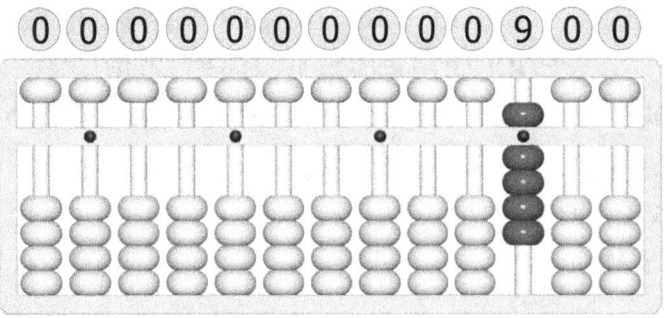

0 0 0 0 0 0 0 0 0 0 9 0 0

13 12 11 10 9 8 7 6 5 4 3 2 1

We will **register 9**

9
- Column 3, register 1 upper bead and 4 lower beads

The abacus reads 9

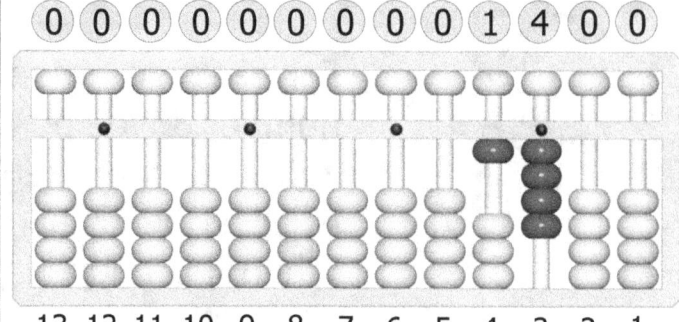

0 0 0 0 0 0 0 0 0 1 4 0 0

13 12 11 10 9 8 7 6 5 4 3 2 1

We will now **add 5**

There are not enough beads in column 3 to add 5, so think **5=10-5**, so remove 5 from column 3 then add 10 to column 4

-5
- Column 3, unregister 1 upper bead to subtract 5

+10
- Column 4, register 1 lower bead to add 10

The abacus result is 14

Example: 156 + 263

0 0 0 0 0 0 0 0 1 5 6 0 0

13 12 11 10 9 8 7 6 5 4 3 2 1

We will **register 156**

100
- Column 5, register 1 lower bead

50
- Column 4, register 1 upper bead

6
- Column 3, register 1 upper bead and 1 lower bead

The abacus reads 156

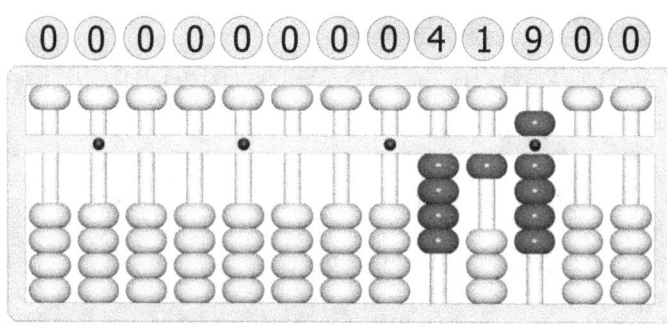

0 0 0 0 0 0 0 0 4 1 9 0 0

13 12 11 10 9 8 7 6 5 4 3 2 1

We will **add 263**

+200
- Column 5, register 2 lower beads

There are not enough beads in column 4 to register 6 more (to add 60), so think **6=10-4**

-40
- Column 4, unregister 1 upper bead (-50) and register 1 lower bead (+10)

+100
- Column 5, register 1 lower bead (Total from columns 4 & 5 is 100-50+10=60)

+3
- Column 3, register 3 lower beads

The abacus result is 419

Addition when registering and unregistering in the same column

Sometimes we need to unregister and register in the same column.
For example if we need to add 3 beads to an already registered 4 lower beads to make 7, we need to register 1 upper and unregister 2 lower (+5-2=3).
Here are some examples.

Example: 204 + 173

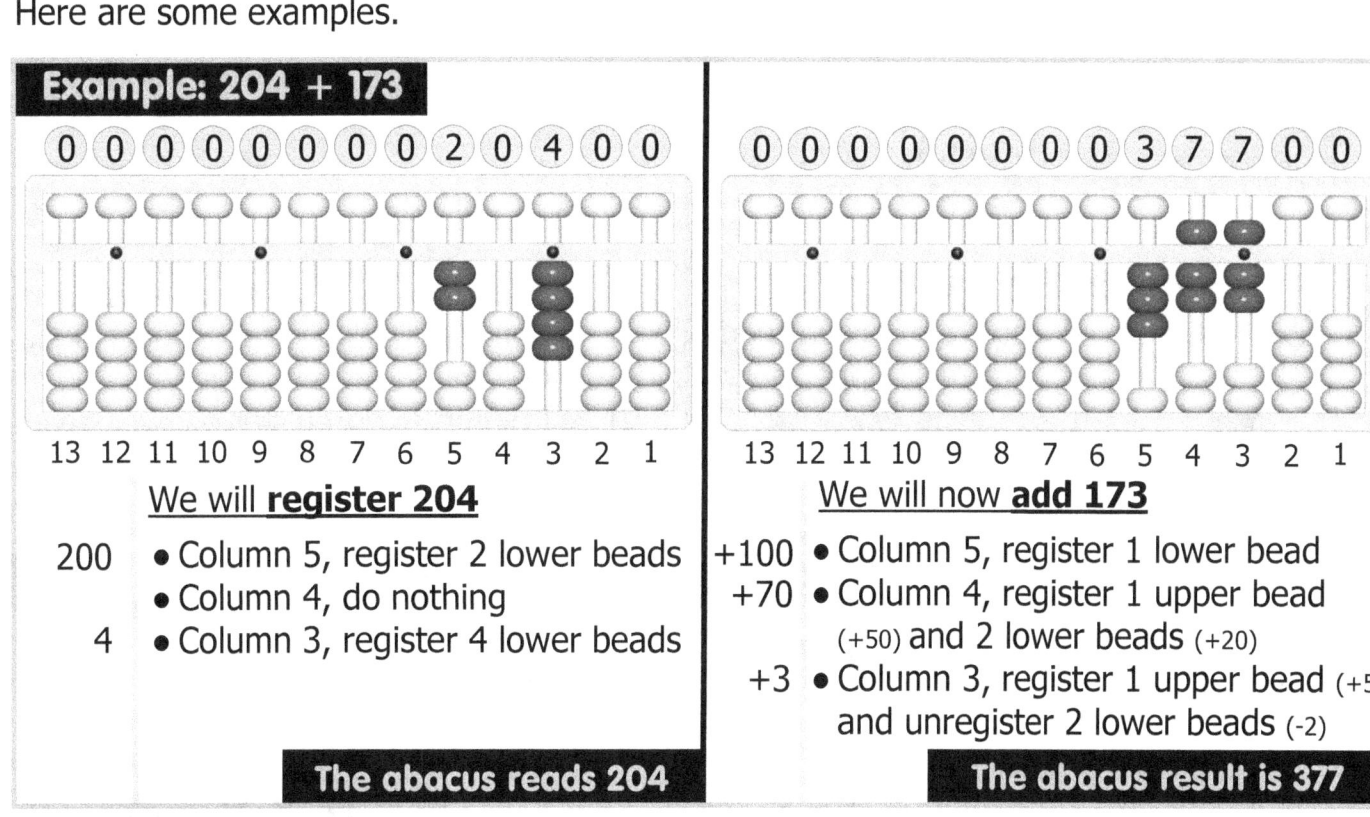

We will register 204

200
- Column 5, register 2 lower beads
- Column 4, do nothing

4
- Column 3, register 4 lower beads

The abacus reads 204

We will now add 173

+100
- Column 5, register 1 lower bead

+70
- Column 4, register 1 upper bead (+50) and 2 lower beads (+20)

+3
- Column 3, register 1 upper bead (+5) and unregister 2 lower beads (-2)

The abacus result is 377

Example: 6421 + 425

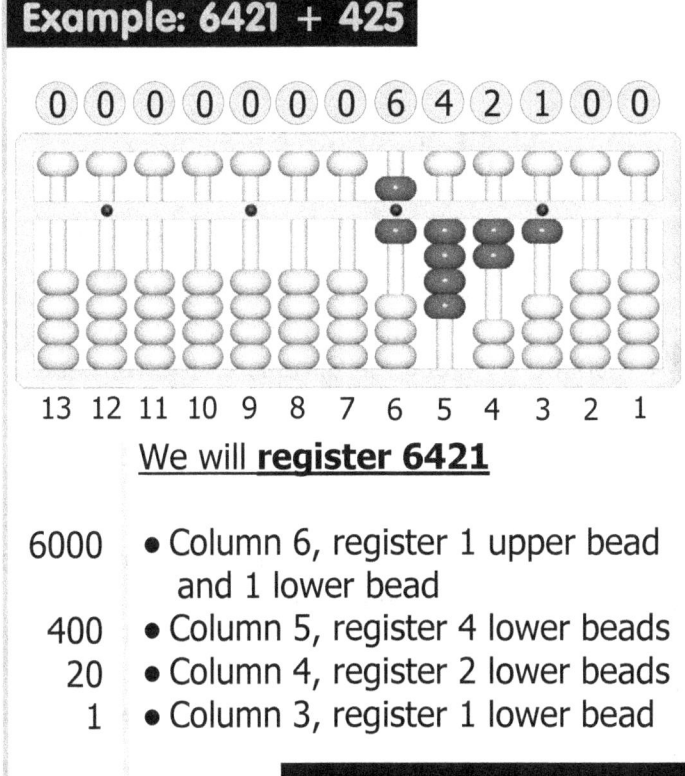
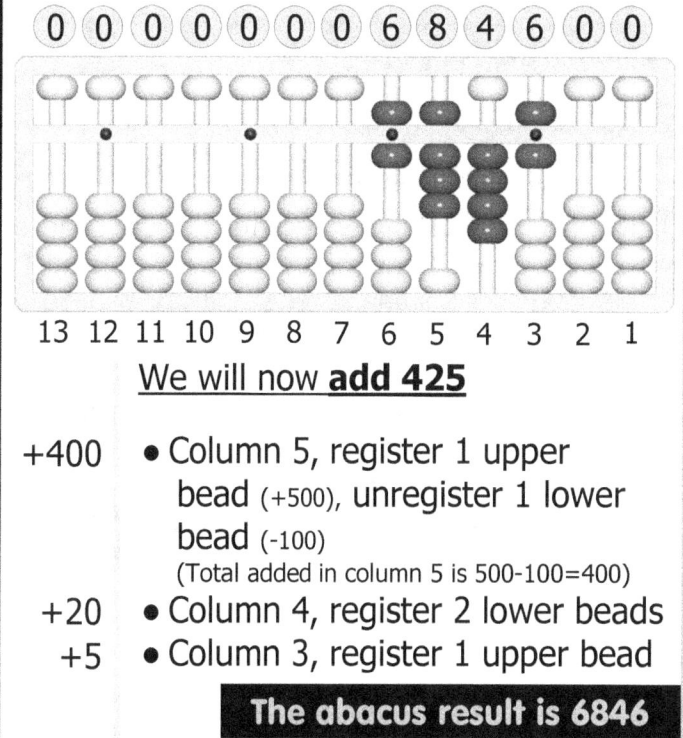

We will register 6421

6000
- Column 6, register 1 upper bead and 1 lower bead

400
- Column 5, register 4 lower beads

20
- Column 4, register 2 lower beads

1
- Column 3, register 1 lower bead

The abacus reads 6421

We will now add 425

+400
- Column 5, register 1 upper bead (+500), unregister 1 lower bead (-100)
  (Total added in column 5 is 500-100=400)

+20
- Column 4, register 2 lower beads

+5
- Column 3, register 1 upper bead

The abacus result is 6846

More addition examples (when we don't have enough beads)

Example: 45 + 5

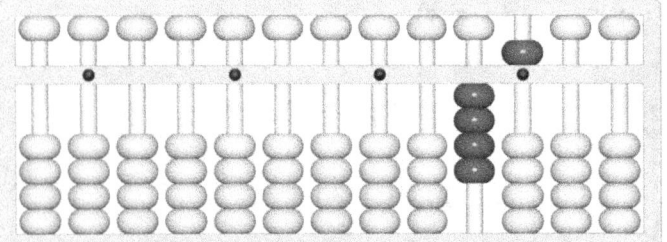

0 0 0 0 0 0 0 0 4 5 0 0

13 12 11 10 9 8 7 6 5 4 3 2 1

We will **register 45**

40 • Column 4, register 4 lower beads
5 • Column 3, register 1 upper bead

LOOK how these make +10
+50-40=+10

The abacus reads 45

---

0 0 0 0 0 0 0 0 5 0 0 0

13 12 11 10 9 8 7 6 5 4 3 2 1

We will **add 5**

There are not enough beads in column 3
to register 5 more (to add 5), so think **5=10-5**

-5 • Column 3, unregister 1 upper bead
+50 • Column 4, register 1 upper bead
-40 • Column 4, unregister 4 lower beads

The abacus result is 50

Example: 5395607 + 2803721

0 0 0 0 5 3 9 5 6 0 7 0 0

13 12 11 10 9 8 7 6 5 4 3 2 1

We will **register 5395607**

5000000 • Column 9, register 1 upper bead
300000 • Column 8, register 3 lower beads
90000 • Column 7, register 1 upper bead
and 4 lower beads
5000 • Column 6, register 1 upper bead
600 • Column 5, register 1 upper bead
and 1 lower bead
• Column 4, do nothing
7 • Column 3, register 1 upper bead
and 2 lower beads

The abacus reads 5395607

---

0 0 0 0 8 1 9 9 3 2 8 0 0

13 12 11 10 9 8 7 6 5 4 3 2 1

We will **add 2803721**

+2000000 • Column 9, register 2 lower beads

There are not enough beads in column 8 to
register 8 more, so think **8=10-2**
-200000 • Column 8, unregister 2 lower beads
+1000000 • Column 9, register 1 lower bead

• Column 7, do nothing
+3000 • Column 6, register 3 lower beads

There are not enough beads in column 5 to
register 7 more, so think **7=10-3**
-500 • Column 5, unregister 1 upper bead
+200 • Column 5, register 2 lower beads
+1000 • Column 6, register 1 lower bead

+20 • Column 4, register 2 lower beads
+1 • Column 3, register 1 lower bead

The abacus result is 8199328

More addition examples (when we don't have enough beads)

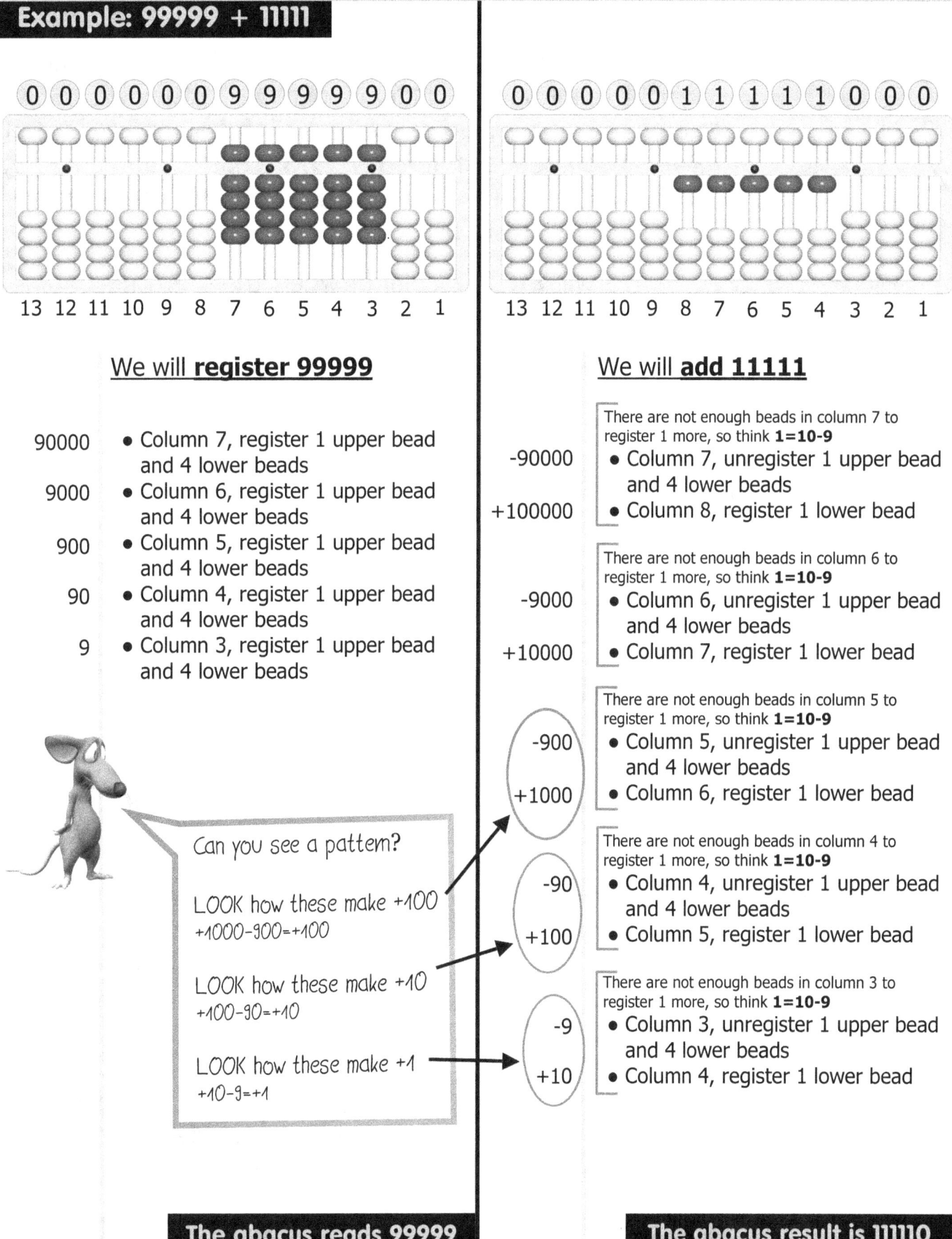

Example: 99999 + 11111

0	0	0	0	0	0	9	9	9	9	9	0	0

13 12 11 10 9 8 7 6 5 4 3 2 1

0	0	0	0	0	1	1	1	1	1	0	0	0

13 12 11 10 9 8 7 6 5 4 3 2 1

We will **register 99999**

90000 • Column 7, register 1 upper bead and 4 lower beads

9000 • Column 6, register 1 upper bead and 4 lower beads

900 • Column 5, register 1 upper bead and 4 lower beads

90 • Column 4, register 1 upper bead and 4 lower beads

9 • Column 3, register 1 upper bead and 4 lower beads

Can you see a pattern?

LOOK how these make +100
+1000-900=+100

LOOK how these make +10
+100-90=+10

LOOK how these make +1
+10-9=+1

We will **add 11111**

There are not enough beads in column 7 to register 1 more, so think **1=10-9**

-90000 • Column 7, unregister 1 upper bead and 4 lower beads

+100000 • Column 8, register 1 lower bead

There are not enough beads in column 6 to register 1 more, so think **1=10-9**

-9000 • Column 6, unregister 1 upper bead and 4 lower beads

+10000 • Column 7, register 1 lower bead

There are not enough beads in column 5 to register 1 more, so think **1=10-9**

-900 / +1000 • Column 5, unregister 1 upper bead and 4 lower beads
 • Column 6, register 1 lower bead

There are not enough beads in column 4 to register 1 more, so think **1=10-9**

-90 / +100 • Column 4, unregister 1 upper bead and 4 lower beads
 • Column 5, register 1 lower bead

There are not enough beads in column 3 to register 1 more, so think **1=10-9**

-9 / +10 • Column 3, unregister 1 upper bead and 4 lower beads
 • Column 4, register 1 lower bead

The abacus reads 99999

The abacus result is 111110

Skipped columns when adding

Sometimes we have to SKIP a column. I'll explain why below.

We've learnt that when a column doesn't have enough beads left on it to make the addition, we move to the next LEFT column to help. Sometimes the next left column also doesn't have enough beads on it, so we **SKIP** this column and move again to the next left column until you reach a column that has enough beads to use. See below how it works.

⁕ Important!

- We will **SKIP** a column when there are not enough beads to use in that column.
- We will see this symbol when we need to skip a column (move on to the next left column).

- We will **UNREGISTER** all beads in any skipped columns.

Example: 95 + 9

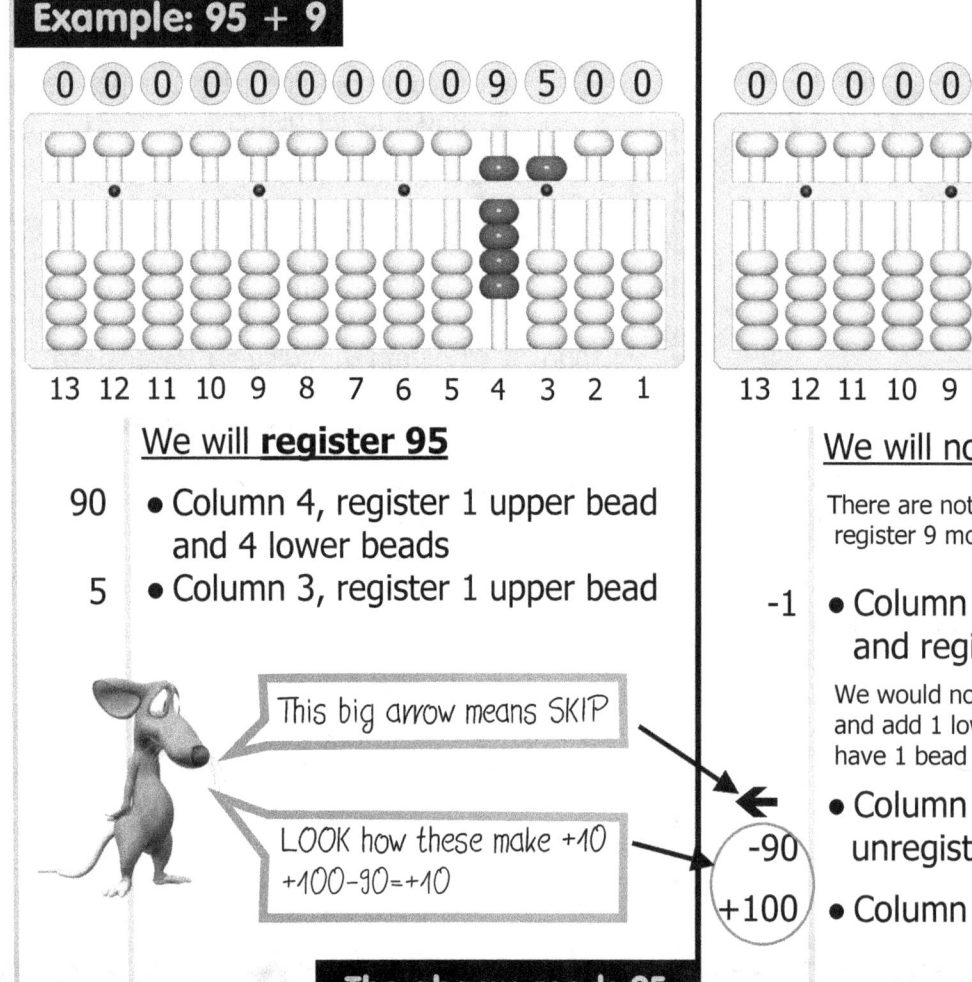

We will **register 95**

90 • Column 4, register 1 upper bead and 4 lower beads

5 • Column 3, register 1 upper bead

This big arrow means SKIP

LOOK how these make +10
+100-90=+10

We will now **add 9**

There are not enough beads in column 3 to register 9 more, so think **9=10-1**

-1 • Column 3, unregister 1 upper bead and register 4 lower beads

We would normally move to the next LEFT column and add 1 lower bead (to add 10) but we don't have 1 bead left to use

-90 • Column 4, **SKIP** this column and unregister all beads

+100 • Column 5, register 1 lower bead

The abacus reads 95

The abacus result is 104

Example: 995 + 9

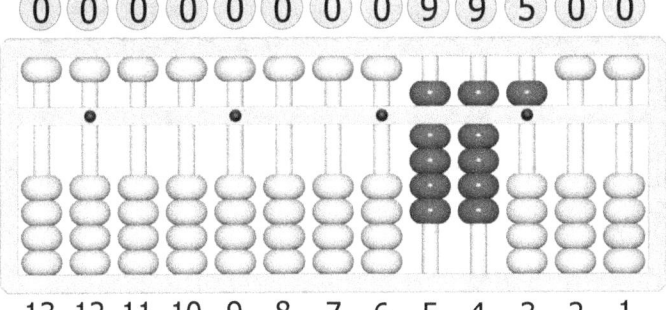

`0 0 0 0 0 0 0 9 9 5 0 0`

13 12 11 10 9 8 7 6 5 4 3 2 1

We will **register 995**

900 • Column 5, register 1 upper bead and 4 lower beads

90 • Column 4, register 1 upper bead and 4 lower beads

5 • Column 3, register 1 upper bead

The abacus reads 995

`0 0 0 0 0 0 0 1 0 0 4 0 0`

13 12 11 10 9 8 7 6 5 4 3 2 1

We will now **add 9**

There are not enough beads in column 3 to register 9 more, so think **9=10-1**

-1 • Column 3, unregister 1 upper bead and register 4 lower beads

← • Column 4, **SKIP** this column and
-90   unregister all beads

← • Column 5, **SKIP** this column and
-900   unregister all beads

+1000 • Column 6, register 1 lower bead

The abacus result is 1004

Example: 9999 + 1

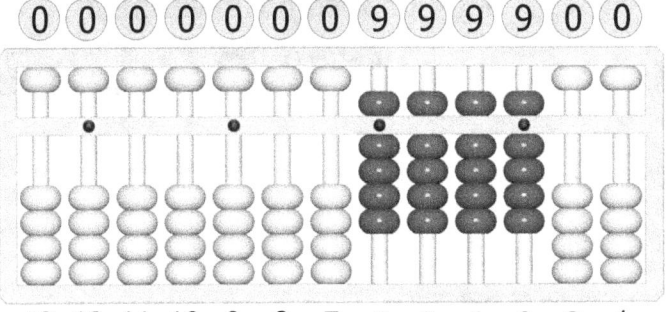

`0 0 0 0 0 0 0 9 9 9 9 0 0`

13 12 11 10 9 8 7 6 5 4 3 2 1

We will **register 9999**

9000 • Column 6, register 1 upper bead and 4 lower beads

900 • Column 5, register 1 upper bead and 4 lower beads

90 • Column 4, register 1 upper bead and 4 lower beads

9 • Column 3, register 1 upper bead and 4 lower beads

The abacus reads 9999

`0 0 0 0 0 0 1 0 0 0 0 0 0`

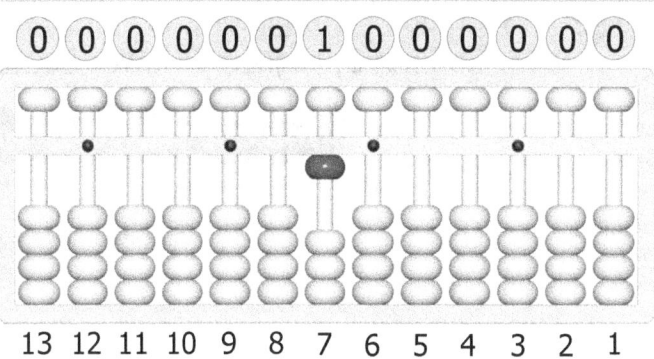

13 12 11 10 9 8 7 6 5 4 3 2 1

We will now **add 1**

There are not enough beads in column 3 to register 1 more, so think **1=10-9**

-9 • Column 3, unregister all beads

← • Column 4, **SKIP** this column and
-90   unregister all beads, move to column 5

← • Column 5, **SKIP** this column and
-900   unregister all beads, move to column 6

← • Column 6, **SKIP** this column and
-9000   unregister all beads, move to column 7

+10000 • Column 7, register 1 lower bead

The abacus result is 10000

Addition of more than two numbers

Sometimes we have to add 3 or more numbers, here's how.

When we add many numbers on the abacus, just find the sum of the first two, then add the next number to that sum.

Keep adding one number to the sum of the previous numbers until all the numbers have been added.

Example: 123 + 254 + 522

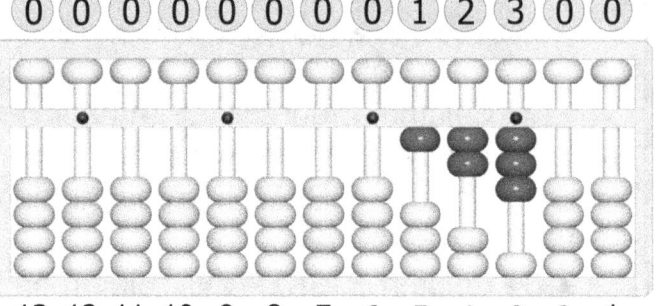

13 12 11 10 9 8 7 6 5 4 3 2 1

We will **register 123**

100	• Column 5, register 1 lower bead
20	• Column 4, register 2 lower beads
3	• Column 3, register 3 lower beads

The abacus reads 123

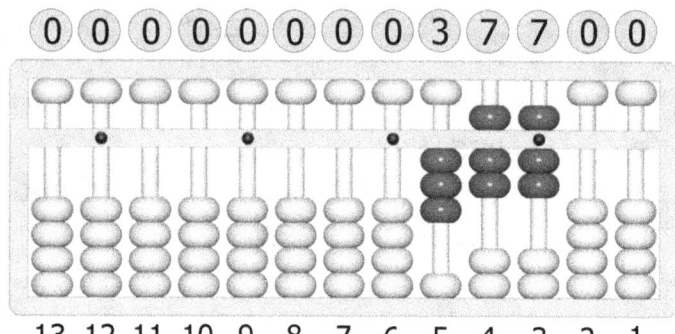

13 12 11 10 9 8 7 6 5 4 3 2 1

We will now **add 254 to 123**

+200	• Column 5, register 2 lower beads
+50	• Column 4, register 1 upper bead
+4	• Column 3, register 1 upper bead and unregister 1 lower bead

The abacus sum is 377

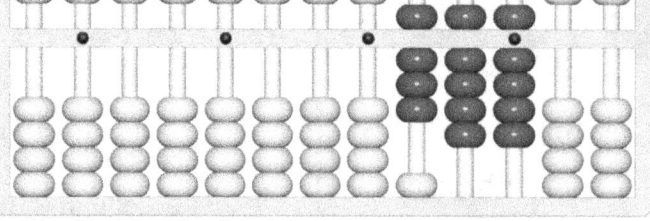

13 12 11 10 9 8 7 6 5 4 3 2 1

We will now **add 522 to the sum 377**

+500	• Column 5, register 1 upper bead
+20	• Column 4, register 2 lower beads
+2	• Column 3, register 2 lower beads

The abacus result is 899

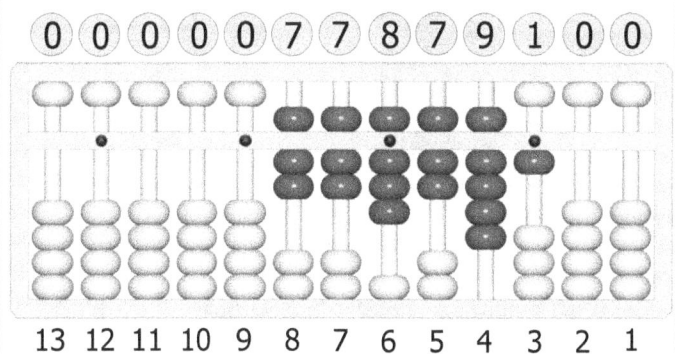

Example: 525631 + 253160 + 1210

0 0 0 0 0 5 2 5 6 3 1 0 0

13 12 11 10 9 8 7 6 5 4 3 2 1

We will **register 525631**

500000	• Column 8, register 1 upper bead
20000	• Column 7, register 2 lower beads
5000	• Column 6, register 1 upper bead
600	• Column 5, register 1 upper bead and 1 lower bead
30	• Column 4, register 3 lower beads
1	• Column 3, register 1 lower bead

The abacus reads 525631

0 0 0 0 0 7 7 8 7 9 1 0 0

13 12 11 10 9 8 7 6 5 4 3 2 1

We will now **add 253160 to 525631**

+200000	• Column 8, register 2 lower beads
+50000	• Column 7, register 1 upper bead
+3000	• Column 6, register 3 lower beads
+100	• Column 5, register 1 lower bead
+60	• Column 4, register 1 upper bead and 1 lower bead
	• Column 3, do nothing

The abacus sum is 778791

0 0 0 0 0 7 8 0 0 0 1 0 0

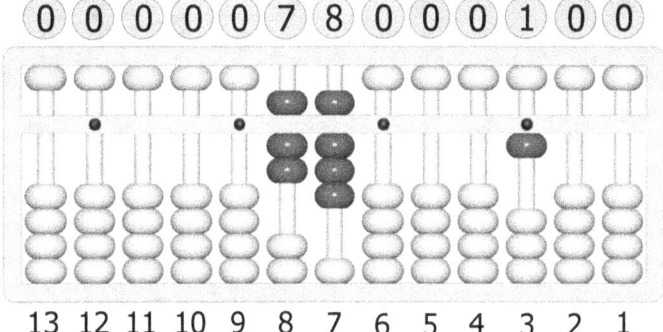

13 12 11 10 9 8 7 6 5 4 3 2 1

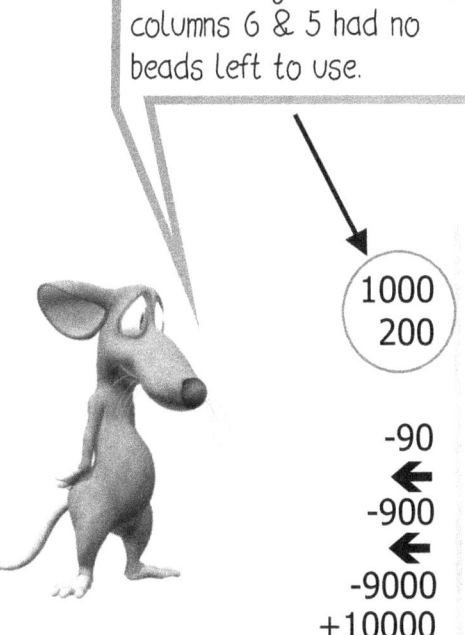

After we registered these, columns 6 & 5 had no beads left to use.

We will now **add 1210 to the sum 778791**

1000	• Column 6, register 1 lower bead
200	• Column 5, register 2 lower beads

> There are not enough beads in column 4 to register 1 more, so think **1=10-9**

-90	• Column 4, unregister all beads
←	• Column 5, **SKIP** this column and unregister all beads, move to column 6
-900	
←	• Column 6, **SKIP** this column and unregister all beads, move to column 7
-9000	
+10000	• Column 7, register 1 lower bead

• Column 3, do nothing

The abacus result is 780001

TEST 2 - Addition

Try to add these numbers on your abacus.

Answers are on page 41 to 43.

 1

$$58 + 45$$

6

$$732689 + 555201$$

2

$$564 + 135$$

7

$$4321543 + 5365$$

 3

$$885 + 9$$

8

$$1535 + 252 + 22$$

4

$$5253 + 3231$$

9

$$135254 + 2560 + 125 + 52151$$

5

$$55227 + 11111$$

10

$$12345678 + 10000009 + 91255450$$

Answers to test 2 (on page 40)

Here are the addition answers.
If you got any wrong, just have another go.

 58 + 45 = **103**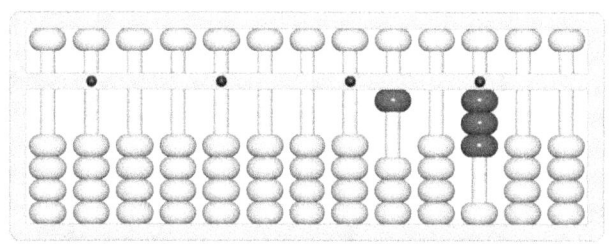

 564 + 135 = **699**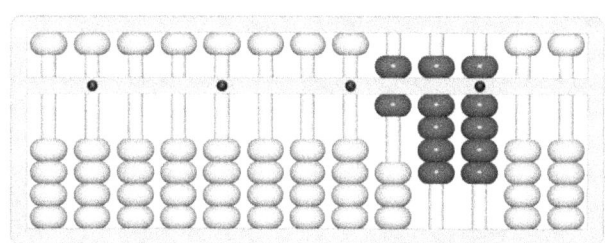

 885 + 9 = **894**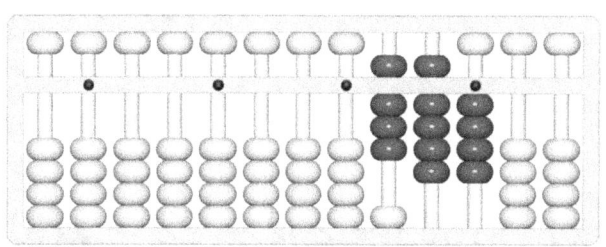

 5253 + 3231 = **8484**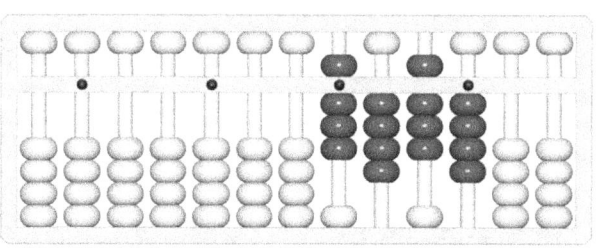

 55227 + 11111 = **66338**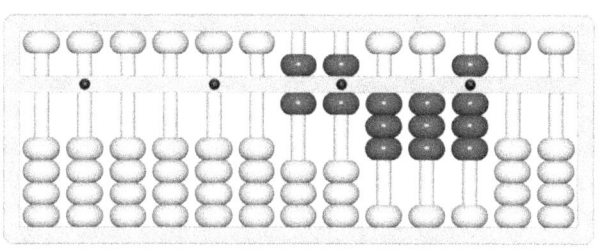

Answers to test 2 (on page 40)

 732689 + 555201 = **1287890**

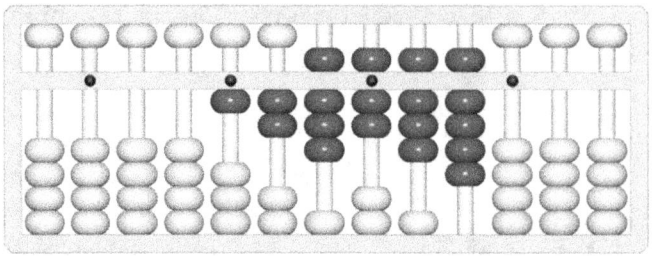

 4321543 + 5365 = **4326908**

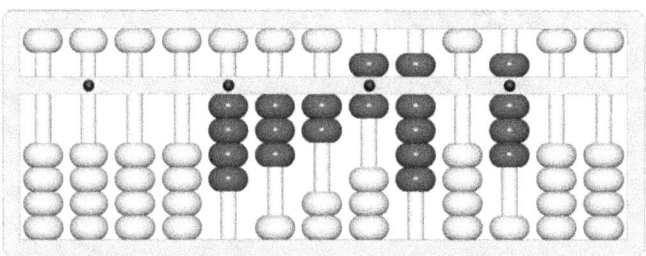

 1535 + 252 + 22 = **1809**

1535 + 252 = **1787**   1787 + 22 = **1809**

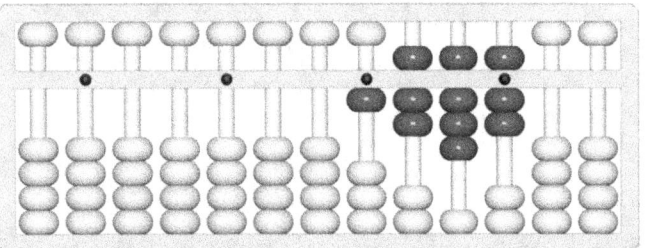 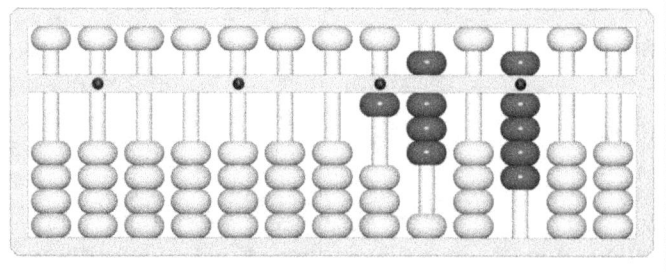

Answers to test 2 (on page 40)

 135254 + 2560 + 125 + 52151 = **190090**

135254 + 2560 = **137814**   137814 + 125 = **137939**

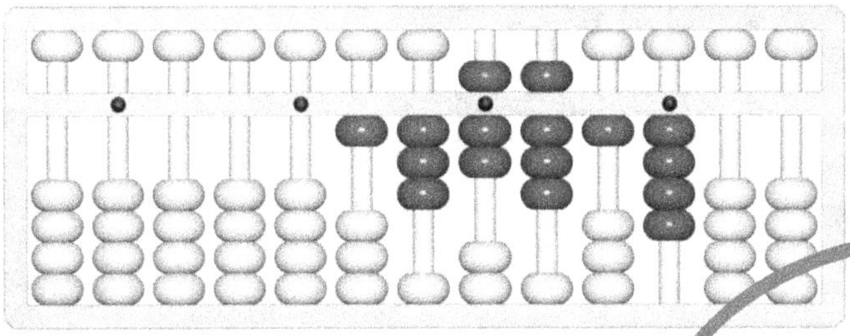

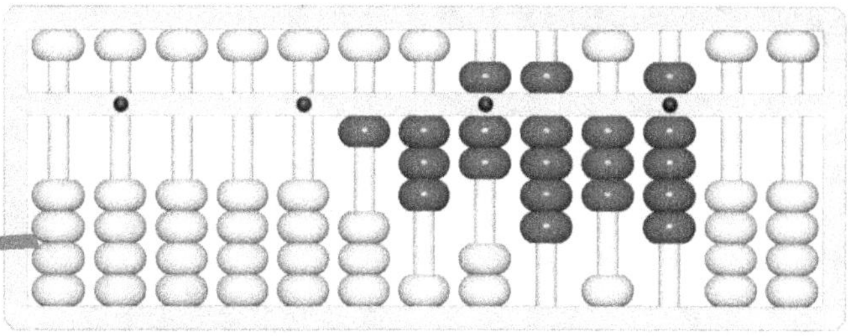

137939 + 52151 = **190090**

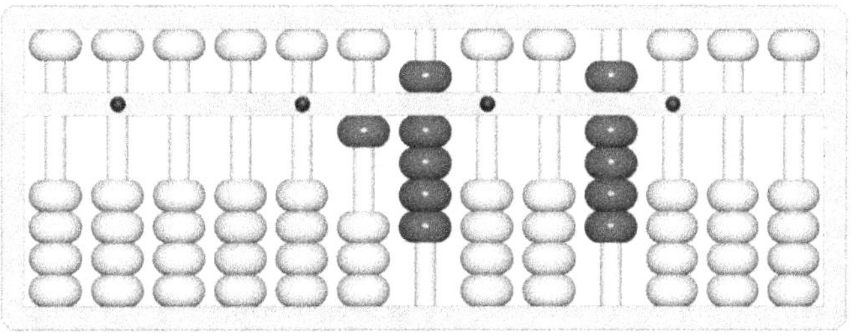

10 12345678 + 10000009 + 91255450 = **113601137**

12345678 + 10000009 = **22345687**   22345687 + 91255450 = **113601137**

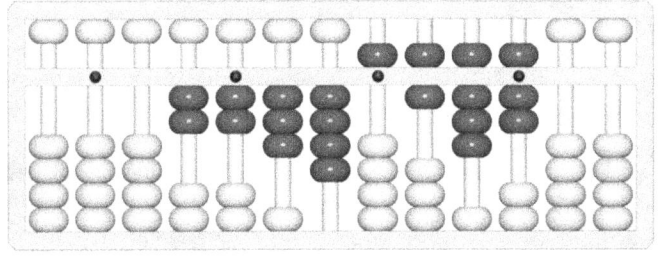
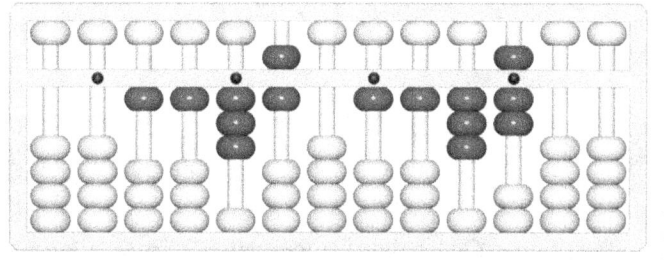

Addition of decimal numbers

A decimal number is a number that contains a decimal point, like 0.5 and it is a part of a whole number.

When we put a decimal number on the abacus, remember we use the dot on the beam to mark our ones column, so any number to the right of the dot (columns 1 & 2) will be our decimals.
Look at these examples:

Example: 129.6

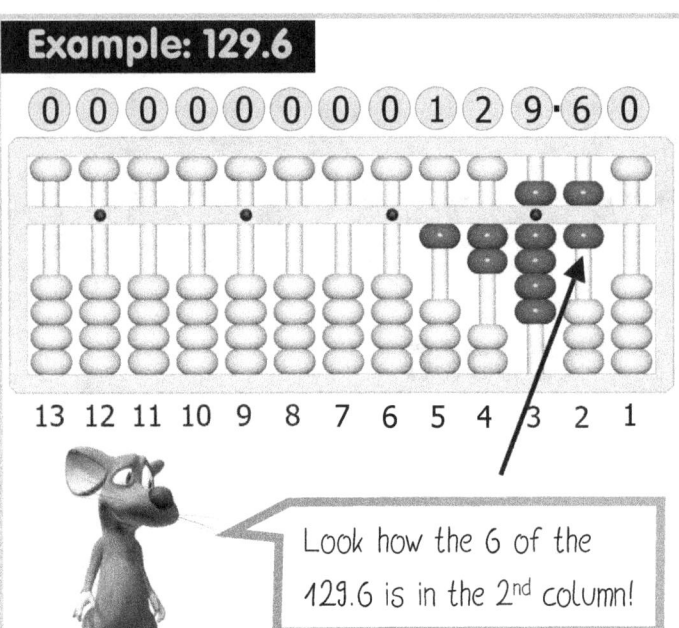

| 0 | 0 | 0 | 0 | 0 | 0 | 0 | 0 | 1 | 2 | 9 | ·6 | 0 |

13 12 11 10 9 8 7 6 5 4 3 2 1

Look how the 6 of the 129.6 is in the 2nd column!

Example: 0.12

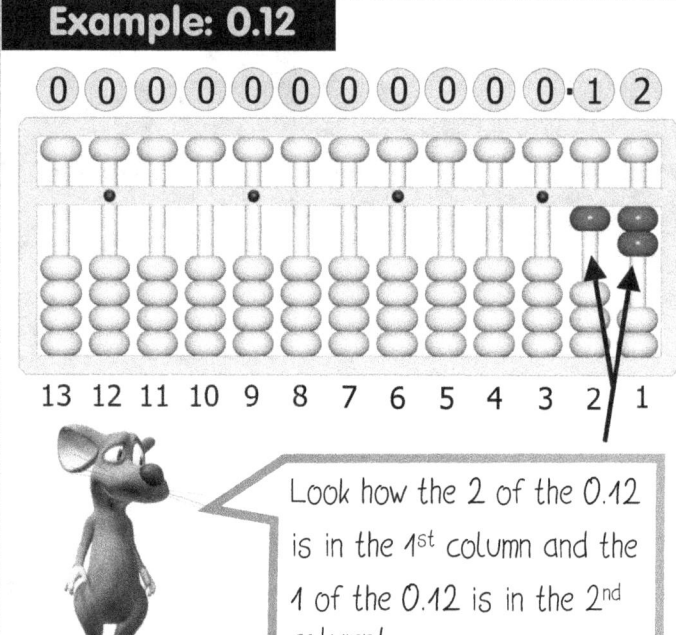

| 0 | 0 | 0 | 0 | 0 | 0 | 0 | 0 | 0 | 0 | 0 | ·1 | 2 |

13 12 11 10 9 8 7 6 5 4 3 2 1

Look how the 2 of the 0.12 is in the 1st column and the 1 of the 0.12 is in the 2nd column!

Example: 32.13 + 12.2

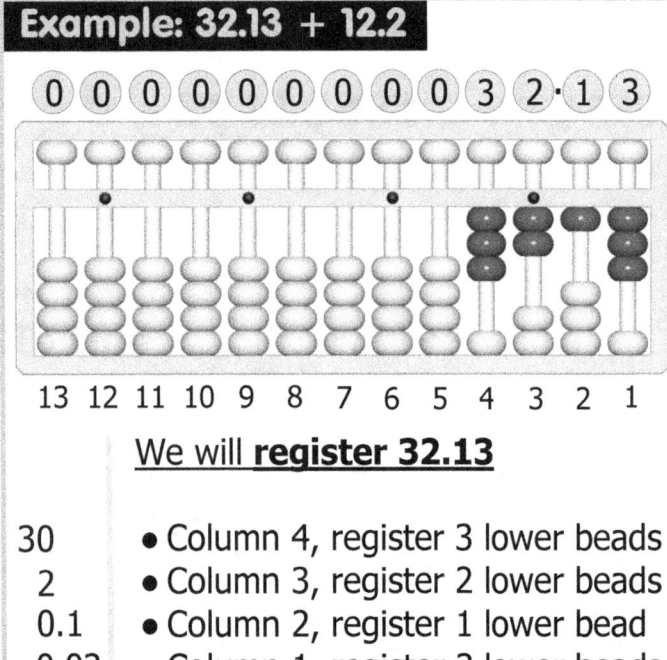

| 0 | 0 | 0 | 0 | 0 | 0 | 0 | 0 | 0 | 3 | 2 | ·1 | 3 |

13 12 11 10 9 8 7 6 5 4 3 2 1

We will **register 32.13**

30	• Column 4, register 3 lower beads
2	• Column 3, register 2 lower beads
0.1	• Column 2, register 1 lower bead
0.03	• Column 1, register 3 lower beads

The abacus reads 32.13

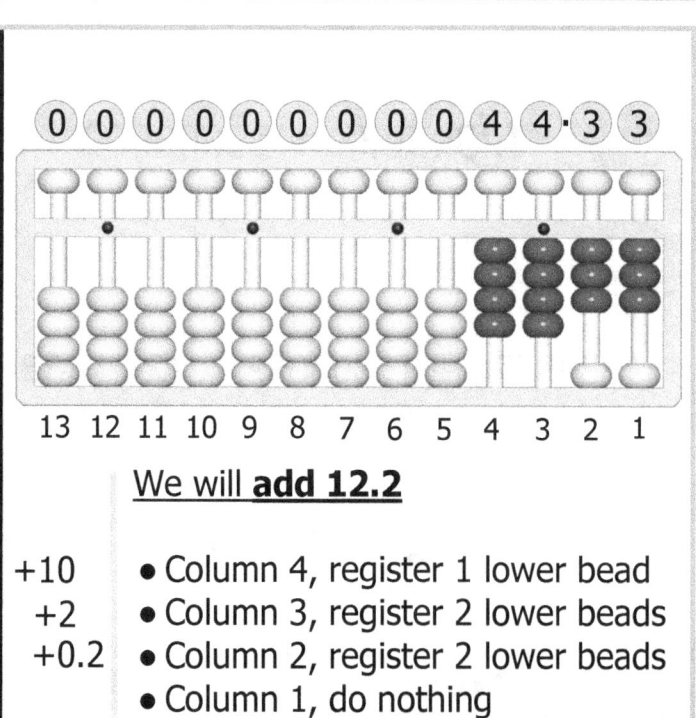

| 0 | 0 | 0 | 0 | 0 | 0 | 0 | 0 | 0 | 4 | 4 | ·3 | 3 |

13 12 11 10 9 8 7 6 5 4 3 2 1

We will **add 12.2**

+10	• Column 4, register 1 lower bead
+2	• Column 3, register 2 lower beads
+0.2	• Column 2, register 2 lower beads
	• Column 1, do nothing

The abacus result is 44.33

Example: 64.37 + 9.32

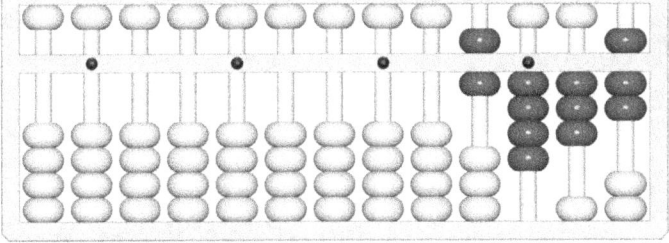

0 0 0 0 0 0 0 0 0 6 4·3 7

13 12 11 10 9 8 7 6 5 4 3 2 1

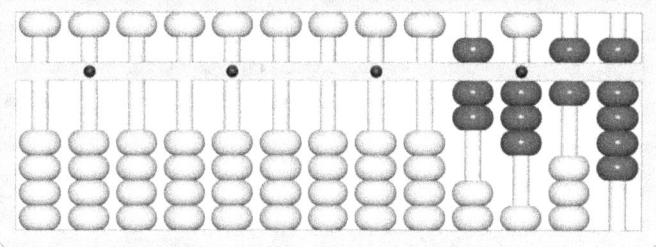

0 0 0 0 0 0 0 0 0 7 3·6 9

13 12 11 10 9 8 7 6 5 4 3 2 1

We will **register 64.37**

60 • Column 4, register 1 upper bead and 1 lower bead

4 • Column 3, register 4 lower beads

0.3 • Column 2, register 3 lower beads

0.07 • Column 1, register 1 upper bead and 2 lower beads

The abacus reads 64.37

We will now **add 9.32**

There are not enough beads in column 3 to register 9 more, so think **9=10-1**

-1 • Column 3, unregister 1 lower bead

+10 • Column 4, register 1 lower bead

+0.3 • Column 2, register 1 upper bead and unregister 2 lower beads

+0.02 • Column 1, register 2 lower beads

The abacus result is 73.69

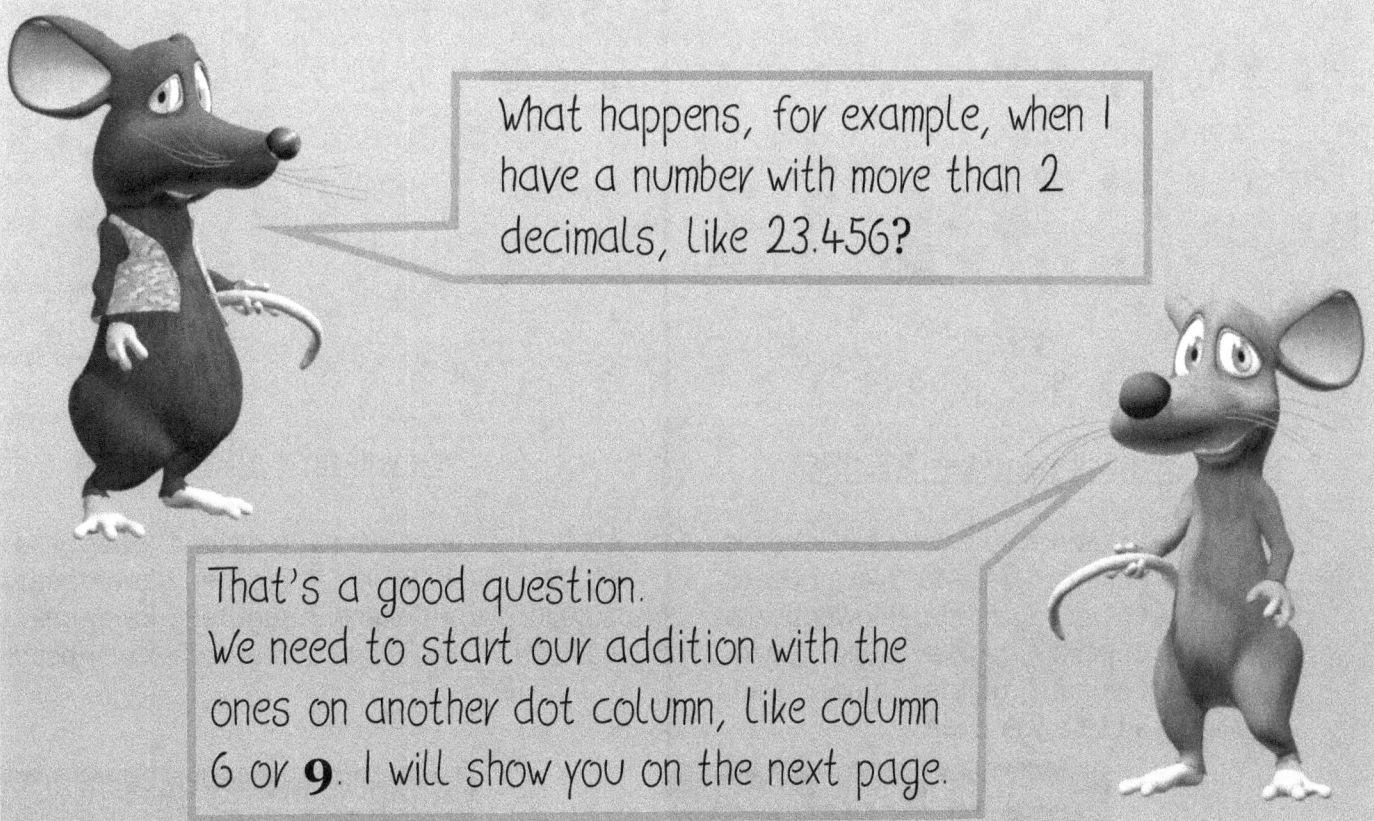

What happens, for example, when I have a number with more than 2 decimals, like 23.456?

That's a good question.
We need to start our addition with the ones on another dot column, like column 6 or **9**. I will show you on the next page.

Adding decimal numbers with more than two decimals

Adding decimal numbers with more than two decimals, isn't so hard to do.

We need to start by registering the first number with the ones on column 6 or 9. In this example 23.456, the ones column can be column 6 (the next dotted column).
So we have to put the 3 of 23.456 on column 6.

Start the 'ones column' here (column 6) instead of the usual column 3. This gives us more columns for our decimals.

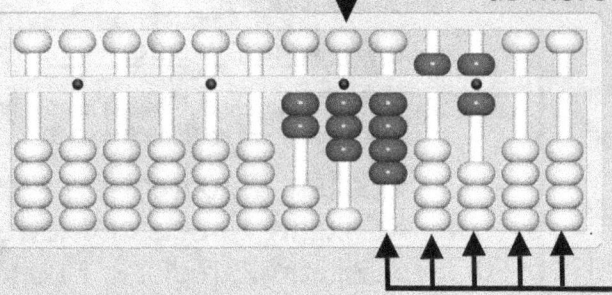

All of these columns can now be used for decimals.

Example: 23.456 + 0.1234

Column 6 is our ones column

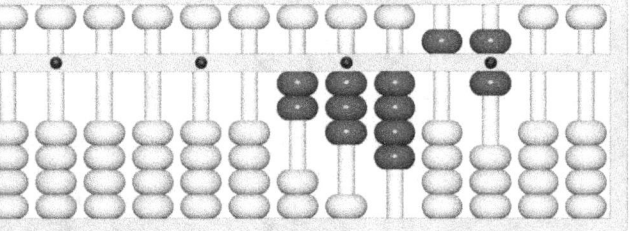

| 0 | 0 | 0 | 0 | 0 | 0 | 2 | 3 · 4 | 5 | 6 | 0 | 0 |

13 12 11 10 9 8 7 6 5 4 3 2 1

We will **register 23.456**

20	• Column 7, register 2 lower beads
3	• Column 6, register 3 lower beads
0.4	• Column 5, register 4 lower beads
0.05	• Column 4, register 1 upper bead
0.006	• Column 3, register 1 upper bead and 1 lower bead

The abacus reads 23.456

Remember to read the answer on the abacus with column 6 as your ones column. The answer is 23.5794 and NOT 23579.4

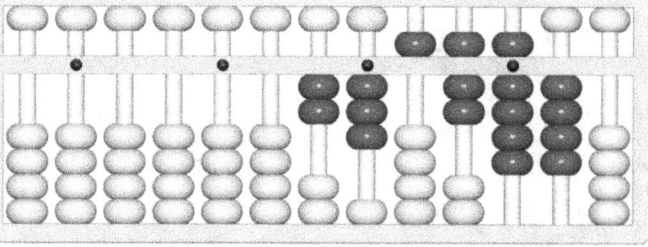

| 0 | 0 | 0 | 0 | 0 | 0 | 2 | 3 · 5 | 7 | 9 | 4 | 0 |

13 12 11 10 9 8 7 6 5 4 3 2 1

We will now **add 0.1234**

+0.1	• Column 5, register 1 lower bead
+0.02	• Column 4, register 2 lower beads
+0.030	• Column 3, register 3 lower beads
+0.0004	• Column 2, register 4 lower beads

The abacus result is 23.5794

TEST 3 - Addition of decimal numbers

Try to add these decimal numbers on your abacus, using column 6 as the ones column.

Use column 6 as the ones column for these questions

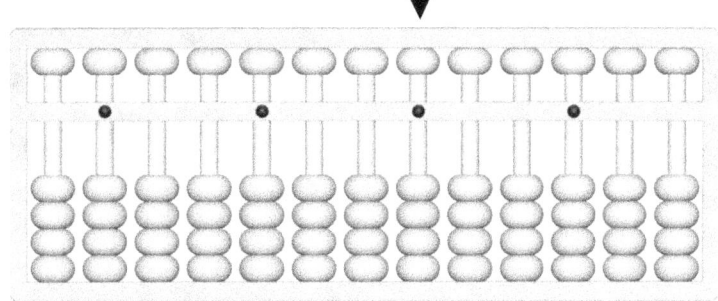

Answers are on page 48.

 1

62.371 + 2.14

 4

2.134 + 0.12345

 2

8.3 + 1.423

5

402.617 + 81.7

 3

0.999 + 0.111

Answers to test 3 (on page 47)

Here are the addition of decimal numbers answers.
If you got any wrong, just have another go.

62.371 + 2.14 = **64.511**

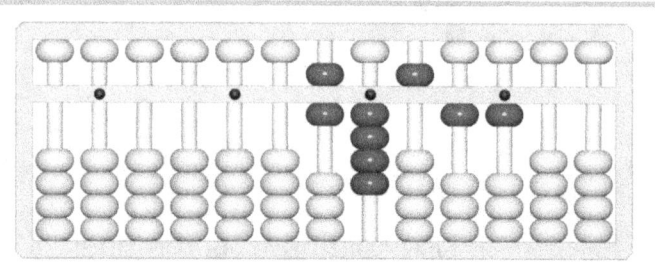

8.3 + 1.423 = **9.723**

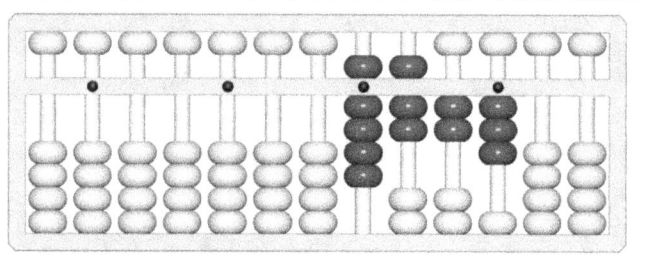

0.999 + 0.111 = **1.11**

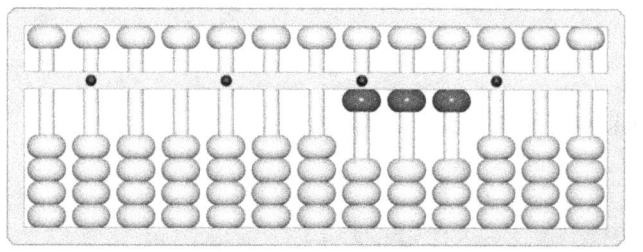

2.134 + 0.12345 = **2.25745**

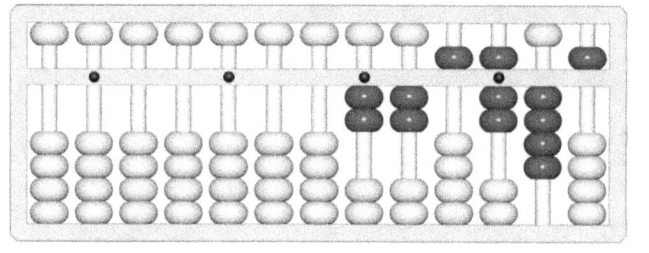

402.617 + 81.7 = **484.317**

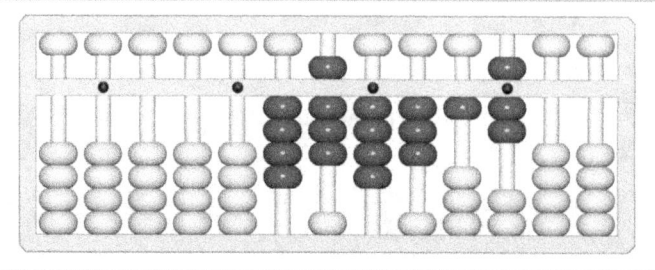

SUBTRACTION

Subtraction is taking one number away from another to find the difference.

Subtraction - things to remember:
- Register your numbers from left to right, just the same as we did with addition, for example:
  for number 231 register the 2 first, 3 second and 1 last.
- Each digit must be registered in the correct column, for example with 231 the 2 is for column 5 (hundredths column), the 3 for column 4 (tens column) and the 1 for column 3 (ones column), just like we did with addition.

Example: 432 - 221

⓪ ⓪ ⓪ ⓪ ⓪ ⓪ ⓪ ⓪ ④ ③ ② ⓪ ⓪

13 12 11 10 9 8 7 6 5 4 3 2 1

We will **register 432**

400 • Column 5, register 4 lower beads
30 • Column 4, register 3 lower beads
2 • Column 3, register 2 lower beads

The abacus reads 432

⓪ ⓪ ⓪ ⓪ ⓪ ⓪ ⓪ ⓪ ② ① ① ⓪ ⓪

13 12 11 10 9 8 7 6 5 4 3 2 1

We will now **subtract 221** from 432

-200 • Column 5, unregister 2 lower beads to subtract 200
-20 • Column 4, unregister 2 lower beads to subtract 20
-1 • Column 3, unregister 1 lower bead to subtract 1

The abacus result is 211

These columns are useful to see the amount that you are subtracting.
For example:
    30 means that you have just registered 30
-200 means that you have just subtracted 200

Subtracting numbers that have different amounts of digits

For example, when subtracting **4234 - 21** we see that 4234 has 4 digits and 21 only has 2.

Register the number that has the largest amount of digits, in this case it is 4234.

Next, subtract the number with the smallest amount of digits, in this example the 21, from the largest digit number.

More subtraction examples

Example: 4234 - 21

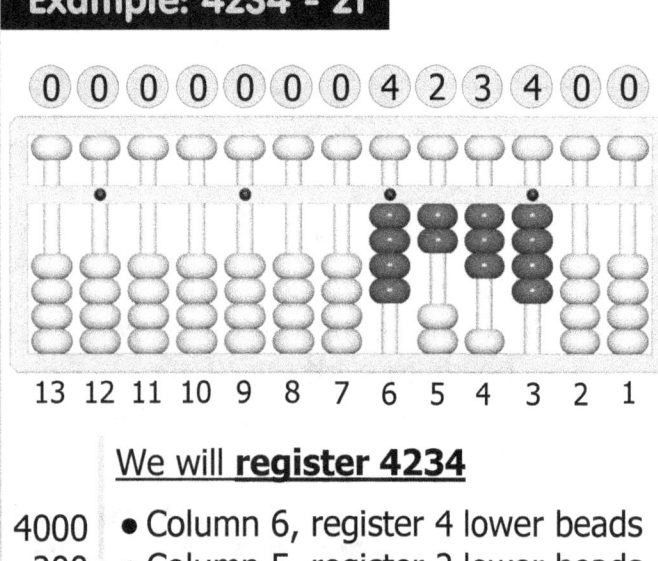

0 0 0 0 0 0 4 2 3 4 0 0

13 12 11 10 9 8 7 6 5 4 3 2 1

<u>We will **register 4234**</u>

4000	• Column 6, register 4 lower beads
200	• Column 5, register 2 lower beads
30	• Column 4, register 3 lower beads
4	• Column 3, register 4 lower beads

Register the number with the most digits first!

The abacus reads 4234

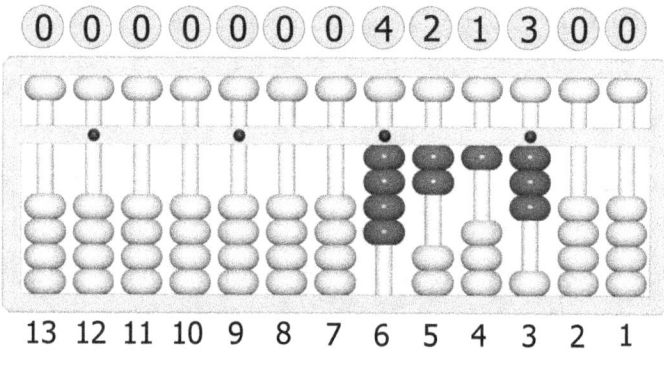

0 0 0 0 0 0 4 2 1 3 0 0

13 12 11 10 9 8 7 6 5 4 3 2 1

<u>We will now **subtract 21**</u>

-20	• Column 4, unregister 2 lower beads to subtract 20
-1	• Column 3, unregister 1 lower bead to subtract 1

The abacus result is 4213

More subtraction examples

Example: 9 - 5

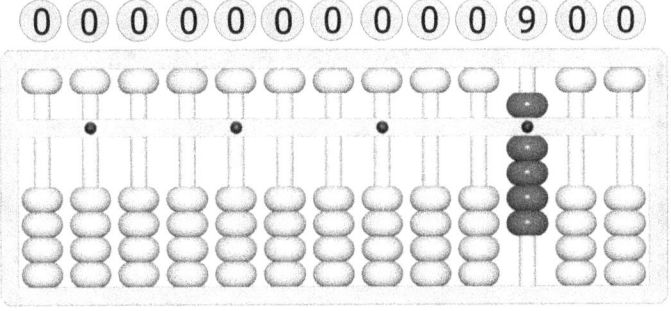

13 12 11 10 9 8 7 6 5 4 3 2 1

We will **register 9**

9 • Column 3, register 1 upper bead and 4 lower beads

The abacus reads 9

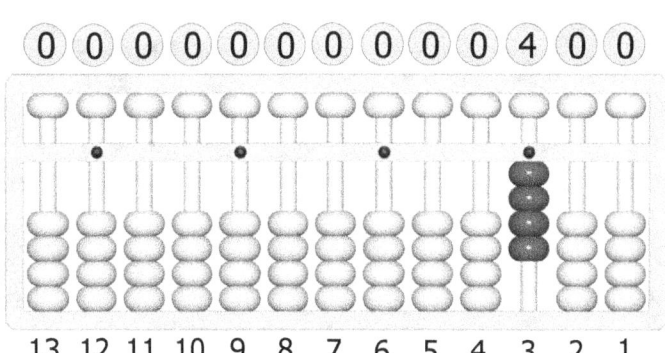

13 12 11 10 9 8 7 6 5 4 3 2 1

We will now **subtract 5**

-5 • Column 3, unregister 1 upper bead

The abacus result is 4

Example: 87 - 21

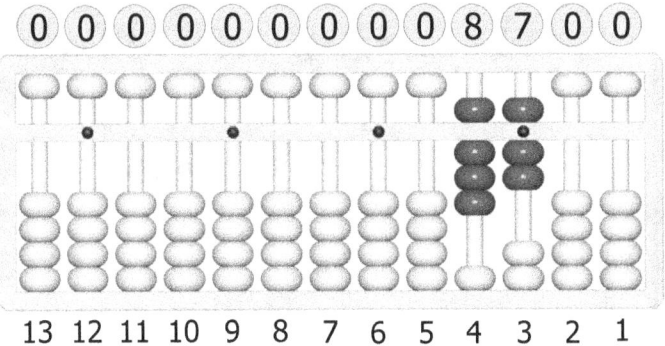

13 12 11 10 9 8 7 6 5 4 3 2 1

We will **register 87**

80 • Column 4, register 1 upper bead and 3 lower beads

7 • Column 3, register 1 upper bead and 2 lower beads

The abacus reads 87

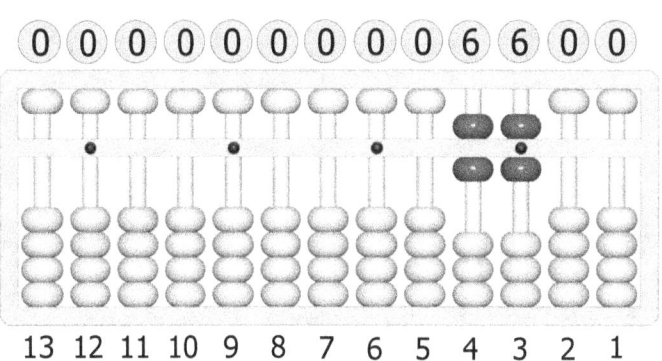

13 12 11 10 9 8 7 6 5 4 3 2 1

We will **subtract 21**

-20 • Column 4, unregister 2 lower beads

-1 • Column 3, unregister 1 lower bead

The abacus result is 66

More subtraction examples

Example: 987 - 671

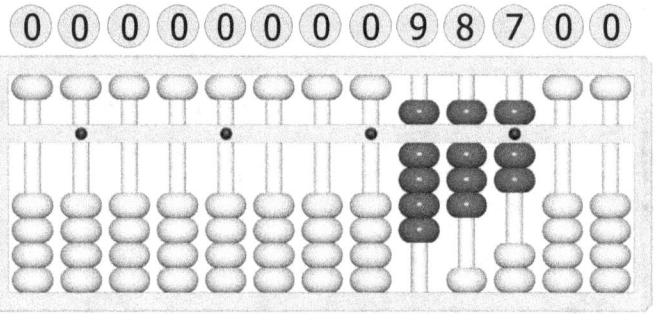

We will **register 987**

900 • Column 5, register 1 upper bead and 4 lower beads

80 • Column 4, register 1 upper bead and 3 lower beads

7 • Column 3, register 1 upper bead and 2 lower beads

The abacus reads 987

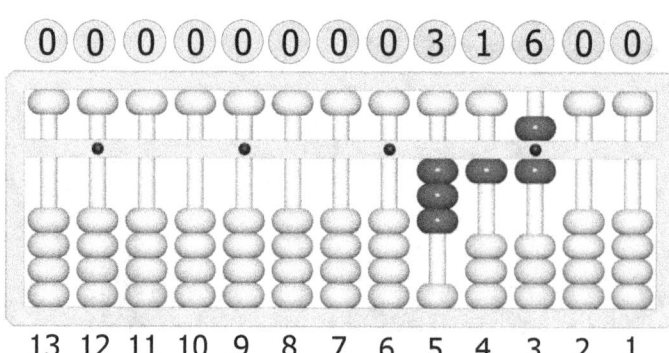

We will now **subtract 671**

-600 • Column 5, unregister 1 upper bead and 1 lower bead
(Total = -500-100=-600)

-70 • Column 4, unregister 1 upper bead and 2 lower beads

-1 • Column 3, unregister 1 lower bead

The abacus result is 316

Example: 6533 - 323

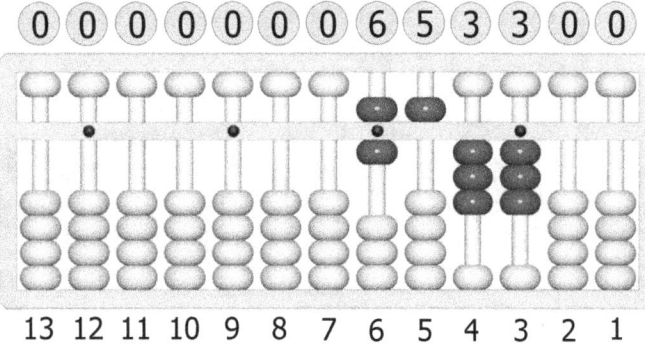

We will **register 6533**

6000 • Column 6, register 1 upper bead and 1 lower bead

500 • Column 5, register 1 upper bead

30 • Column 4, register 3 lower beads

3 • Column 3, register 3 lower beads

The abacus reads 6533

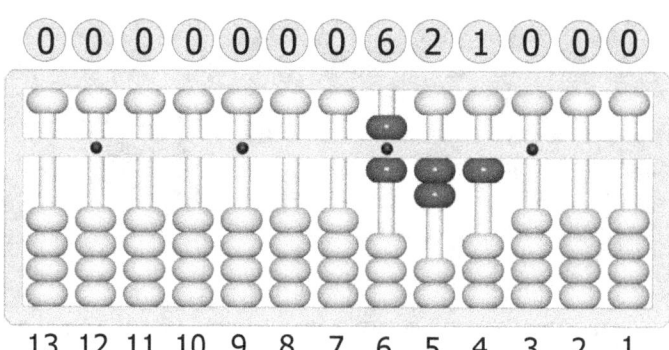

We will now **subtract 323**

-300 • Column 5, unregister 1 upper bead and register 2 lower beads
(Total = -500+200=-300)

-20 • Column 4, unregister 2 lower beads

-3 • Column 3, unregister 3 lower beads

The abacus result is 6210

Example: 76205 - 5203

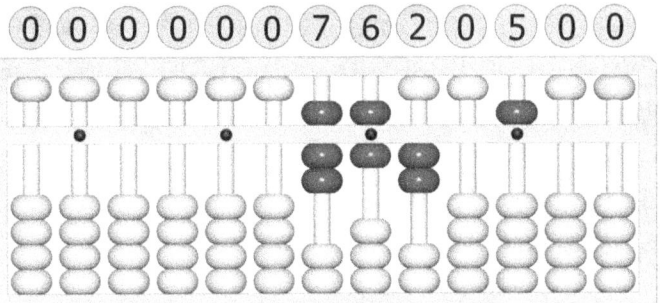

	0	0	0	0	0	0	7	6	2	0	5	0	0
	13	12	11	10	9	8	7	6	5	4	3	2	1

We will **register 76205**

70000	• Column 7, register 1 upper bead and 2 lower beads
6000	• Column 6, register 1 upper bead and 1 lower bead
200	• Column 5, register 2 lower beads
	• Column 4, do nothing
5	• Column 3, register 1 upper bead

The abacus reads 76205

We will now **subtract 5203**

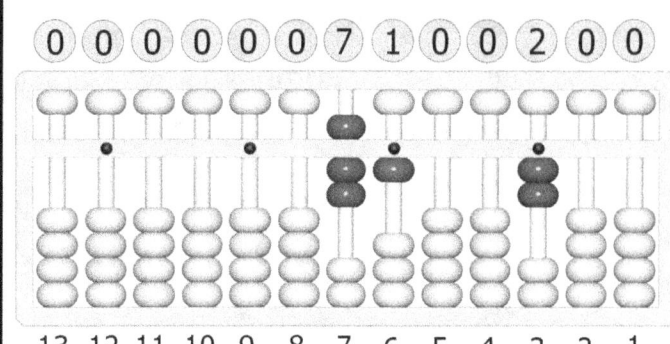

	0	0	0	0	0	0	7	1	0	0	2	0	0
	13	12	11	10	9	8	7	6	5	4	3	2	1

-5000	• Column 6, unregister 1 upper bead
-200	• Column 5, unregister 2 lower beads
	• Column 4, do nothing
-3	• Column 3, unregister 1 upper bead and register 2 lower beads

The abacus result is 71002

Example: 642603284 - 521502064

	0	0	6	4	2	6	0	3	2	8	4	0	0
	13	12	11	10	9	8	7	6	5	4	3	2	1

We will **register 642603284**

600000000	• Column 11, register 1 upper and 1 lower bead
40000000	• Column 10, register 4 lower beads
2000000	• Column 9, register 2 lower beads
600000	• Column 8, register 1 upper and 1 lower bead
	• Column 7, do nothing
3000	• Column 6, register 3 lower beads
200	• Column 5, register 2 lower beads
80	• Column 4, register 1 upper and 3 lower beads
4	• Column 3, register 4 lower beads

The abacus reads 642603284

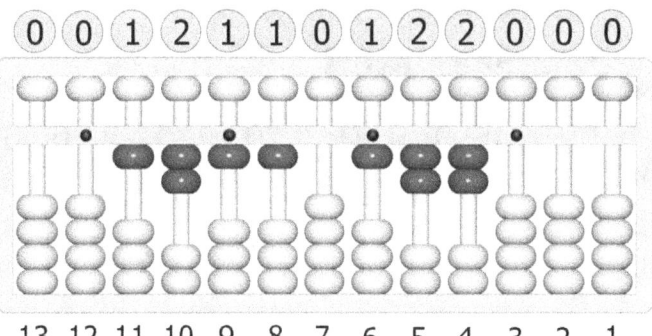

	0	0	1	2	1	1	0	1	2	2	0	0	0
	13	12	11	10	9	8	7	6	5	4	3	2	1

We will **subtract 521502064**

-500000000	• Column 11, unregister 1 upper bead
-20000000	• Column 10, unregister 2 lower beads
-1000000	• Column 9, unregister 1 lower bead
-500000	• Column 8, unregister 1 upper bead
	• Column 7, do nothing
-2000	• Column 6, unregister 2 lower beads
	• Column 5, do nothing
-60	• Column 4, unregister 1 upper and 1 lower bead
-4	• Column 3, unregister 4 lower beads

The abacus result is 121101220

When there are not enough beads in the column for the subtraction

When you don't have enough beads, move to the next LEFT column to help

For example, when you try to subtract 8 from an already registered number 12, you don't have enough beads in the column where the 2 of the 12 is, to do it. You can only unregister a maximum of 9 in each column (4 lower beads and 1 upper bead, -4-5=-9).
When this happens, we need to use the
'Not enough beads list for subtraction'.

```
-1=-10+9
-2=-10+8
-3=-10+7
-4=-10+6
-5=-10+5
-6=-10+4
-7=-10+3
-8=-10+2
-9=-10+1
```

How to use the 'Not enough beads list for subtraction'

Let's say we need to subtract 8 from a column but we don't have enough beads.

Look at the list, **-8=-10+2**

10 is the number to **unregister,** in the next **LEFT** column (1 lower bead).

2 is the number to **register** in our column.

Example: 12 - 8

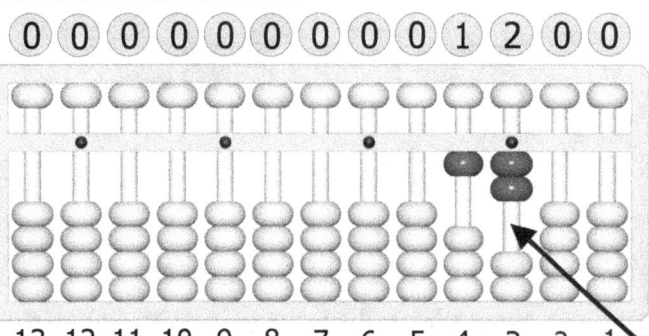

`0 0 0 0 0 0 0 0 0 1 2 0 0`

13 12 11 10 9 8 7 6 5 4 3 2 1

We will **register 12**

10 • Column 4, register 1 lower bead
2 • Column 3, register 2 lower beads

Unregister means move away from the beam (subtract)!

The abacus reads 12

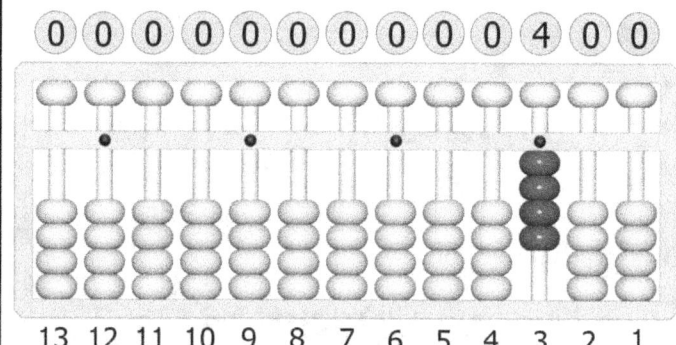

`0 0 0 0 0 0 0 0 0 0 4 0 0`

13 12 11 10 9 8 7 6 5 4 3 2 1

We will now **subtract 8**

There are not enough beads in column 3 to subtract 8, move to the next LEFT column to help.
First we must think **-8=-10+2** see the
'Not enough beads list for subtraction'

-10 • Column 4, **unregister** 1 lower bead to subtract 10
+2 • Column 3, **register** 2 lower beads to add 2 (from -8=-10+2)

The abacus result is 4

More subtraction examples (when we don't have enough beads)

Example: 25 - 16

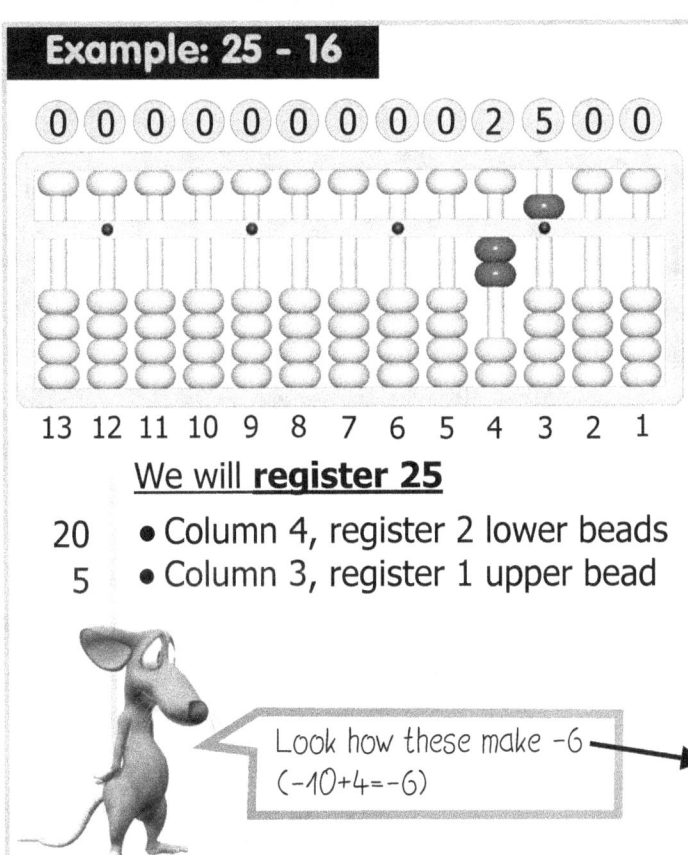

```
0 0 0 0 0 0 0 0 0 2 5 0 0
13 12 11 10 9 8 7 6 5 4 3 2 1
```

We will <u>register 25</u>

20 • Column 4, register 2 lower beads

5 • Column 3, register 1 upper bead

Look how these make -6
(-10+4=-6)

The abacus reads 25

```
0 0 0 0 0 0 0 0 0 0 9 0 0
13 12 11 10 9 8 7 6 5 4 3 2 1
```

We will now <u>subtract 16</u>

-10 • Column 4, unregister 1 lower bead

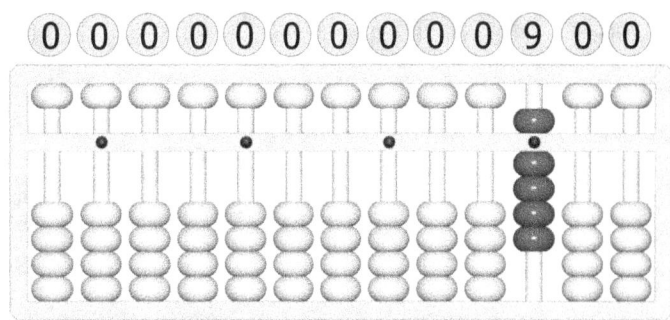

There are not enough beads in column 3 to subtract 6, so move to the next left column to help and think **-6=-10+4**

-10 • Column 4, unregister 1 lower bead to subtract 10

+4 • Column 3, register 4 lower beads to add 4

The abacus result is 9

Example: 10 - 5

```
0 0 0 0 0 0 0 0 0 1 0 0 0
13 12 11 10 9 8 7 6 5 4 3 2 1
```

We will <u>register 10</u>

10 • Column 4, register 1 lower bead

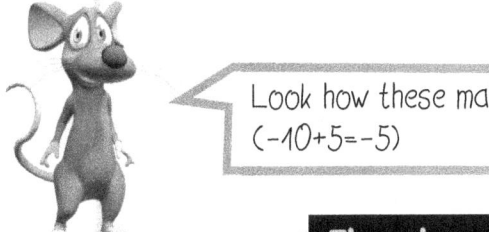

Look how these make -5
(-10+5=-5)

The abacus reads 10

```
0 0 0 0 0 0 0 0 0 0 5 0 0
13 12 11 10 9 8 7 6 5 4 3 2 1
```

We will <u>subtract 5</u>

There are not enough beads in column 3 to subtract 5, so move to the next left column to help and think **-5=-10+5**

-10 • Column 4, unregister 1 lower bead

+5 • Column 3, register 1 upper bead

The abacus result is 5

More subtraction examples (when we don't have enough beads)

Example: 477 - 286

0 0 0 0 0 0 0 0 4 7 7 0 0

13 12 11 10 9 8 7 6 5 4 3 2 1

We will **register 477**

400 • Column 5, register 4 lower beads
70   • Column 4, register 1 upper bead and 2 lower beads
7   • Column 3, register 1 upper bead and 2 lower beads

The abacus reads 477

(top right)

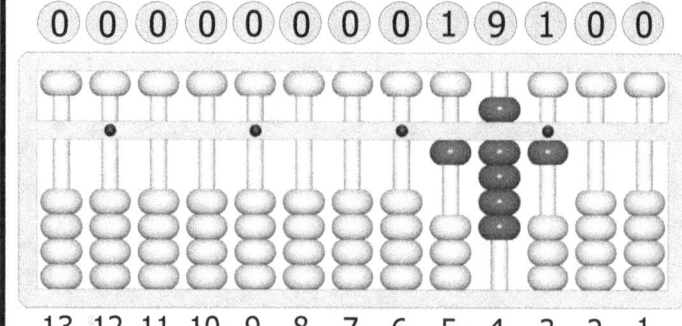

0 0 0 0 0 0 0 0 1 9 1 0 0

13 12 11 10 9 8 7 6 5 4 3 2 1

We will now **subtract 286**

-200 • Column 5, unregister 2 lower beads

> There are not enough beads in column 4 to subtract 8, so move to the next left column to help and think **-8=-10+2**

-100 • Column 5, unregister 1 lower bead
+20   • Column 4, register 2 lower beads

-6   • Column 3, unregister 1 upper and 1 lower bead

The abacus result is 191

Example: 463 - 386

0 0 0 0 0 0 0 0 4 6 3 0 0

13 12 11 10 9 8 7 6 5 4 3 2 1

We will **register 463**

400 • Column 5, register 4 lower beads
60   • Column 4, register 1 upper bead and 1 lower bead
3   • Column 3, register 3 lower beads

Look how these make -80 and these make -6

The abacus reads 463

(bottom right)

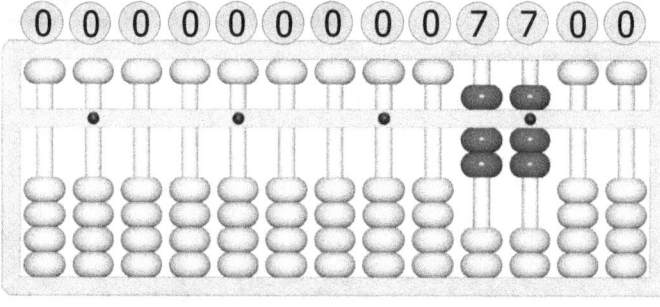

0 0 0 0 0 0 0 0 7 7 0 0

13 12 11 10 9 8 7 6 5 4 3 2 1

We will **subtract 386**

-300 • Column 5, unregister 3 lower beads

> There are not enough beads in column 4 to subtract 8, so move to the next left column to help and think **-8=-10+2**

-100 • Column 5, unregister 1 lower bead
+20   • Column 4, register 2 lower beads

> There are not enough beads in column 3 to subtract 6, so move to the next left column to help and think **-6=-10+4**

-10   • Column 4, unregister 1 lower bead
+4   • Column 3, register 1 upper bead and un-register 1 lower bead (+5-1=+4)

The abacus result is 77

More subtraction examples (when we don't have enough beads)

Example: 6533 - 600

0 0 0 0 0 0 0 6 5 3 3 0 0

13 12 11 10 9 8 7 6 5 4 3 2 1

We will **register 6533**

6000	• Column 6, register 1 upper and 1 lower bead
500	• Column 5, register 1 upper bead
30	• Column 4, register 3 lower beads
3	• Column 3, register 3 lower beads

The abacus reads 6533

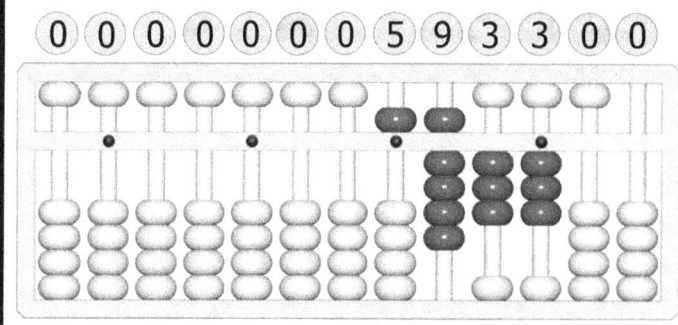

0 0 0 0 0 0 0 5 9 3 3 0 0

13 12 11 10 9 8 7 6 5 4 3 2 1

We will now **subtract 600**

There are not enough beads in column 5 to subtract 6, so move to the next left column to help and think **-6=-10+4**

-1000	• Column 6, unregister 1 lower bead
+400	• Column 5, register 4 lower beads
	• Columns 4 & 3, do nothing

The abacus result is 5933

Example: 6495613 - 5530232

0 0 0 0 6 4 9 5 6 1 3 0 0

13 12 11 10 9 8 7 6 5 4 3 2 1

We will **register 6495613**

6000000	• Column 9, register 1 upper and 1 lower bead
400000	• Column 8, register 4 lower beads
90000	• Column 7, register 1 upper and 4 lower beads
5000	• Column 6, register 1 upper bead
600	• Column 5, register 1 upper and 1 lower bead
10	• Column 4, register 1 lower bead
3	• Column 3, register 3 lower beads

The abacus reads 6495613

---

0 0 0 0 0 9 6 5 3 8 1 0 0

13 12 11 10 9 8 7 6 5 4 3 2 1

We will now **subtract 5530232**

-5000000	• Column 9, unregister 1 upper bead
	Not enough beads in column 8 to subtract 5, so think **-5=-10+5**
-1000000	• Column 9, unregister 1 lower bead
+500000	• Column 8, register 1 upper bead
-30000	• Column 7, unregister 3 lower beads
	• Column 6, do nothing
-200	• Column 5, unregister 1 upper bead and register 3 lower beads
	Not enough beads in column 4 to subtract 3, so think **-3=-10+7**
-100	• Column 5, unregister 1 lower bead
+50	• Column 4, register 1 upper bead
+20	• Column 4, register 2 lower beads
-2	• Column 3, unregister 2 lower beads

The abacus result is 965381

More subtraction examples (when we don't have enough beads)

Example: 100000 - 11111

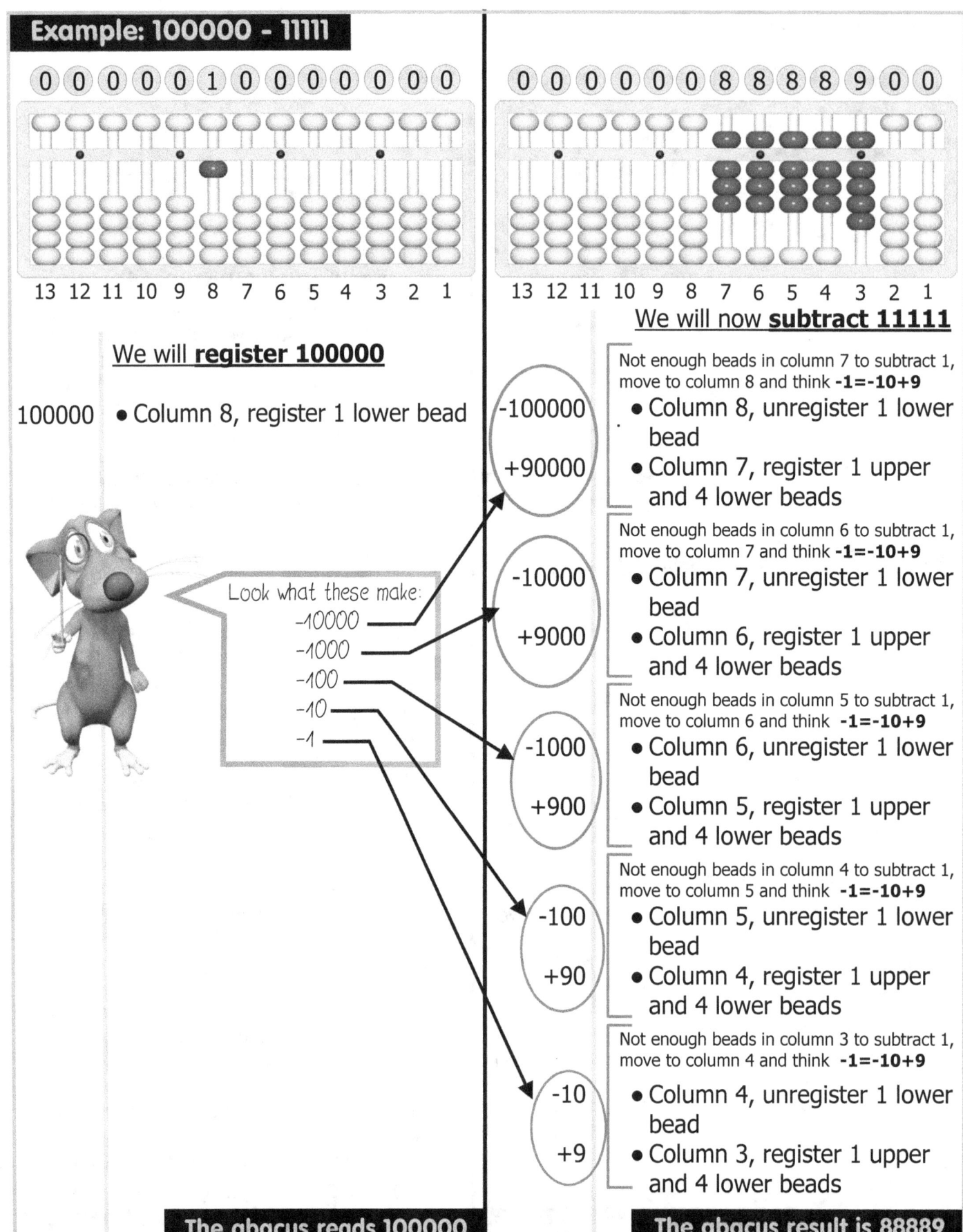

`0 0 0 0 0 1 0 0 0 0 0 0 0`

13 12 11 10 9 8 7 6 5 4 3 2 1

`0 0 0 0 0 0 8 8 8 8 9 0 0`

13 12 11 10 9 8 7 6 5 4 3 2 1

We will **register 100000**

100000 • Column 8, register 1 lower bead

Look what these make:
−10000
−1000
−100
−10
−1

We will now **subtract 11111**

-100000
+90000

Not enough beads in column 7 to subtract 1, move to column 8 and think **-1=-10+9**
• Column 8, unregister 1 lower bead
• Column 7, register 1 upper and 4 lower beads

-10000
+9000

Not enough beads in column 6 to subtract 1, move to column 7 and think **-1=-10+9**
• Column 7, unregister 1 lower bead
• Column 6, register 1 upper and 4 lower beads

-1000
+900

Not enough beads in column 5 to subtract 1, move to column 6 and think **-1=-10+9**
• Column 6, unregister 1 lower bead
• Column 5, register 1 upper and 4 lower beads

-100
+90

Not enough beads in column 4 to subtract 1, move to column 5 and think **-1=-10+9**
• Column 5, unregister 1 lower bead
• Column 4, register 1 upper and 4 lower beads

-10
+9

Not enough beads in column 3 to subtract 1, move to column 4 and think **-1=-10+9**
• Column 4, unregister 1 lower bead
• Column 3, register 1 upper and 4 lower beads

The abacus reads 100000

The abacus result is 88889

Skipped columns when subtracting

Like we did with addition, sometimes with subtraction we have to SKIP a column. I'll explain why below.

When a column doesn't have enough beads left on it to make the subtraction, we move to the next LEFT column to help. Sometimes the next left column also doesn't have enough beads on it, so we **SKIP** this column and move again to the next left column until you reach a column that has enough beads to use. See below how it works.

※ Important!

- We will **SKIP** a column when there are not enough beads to use in that column.

- We will see this symbol when we need to **SKIP** ← a column (move on to the next left column).

- We will see this symbol when we need to **MOVE BACK** → a column (move on to the next right column).

- We will **REGISTER** all beads in any skipped columns (with addition we unregistered, here we do the opposite).

Example: 100 - 5

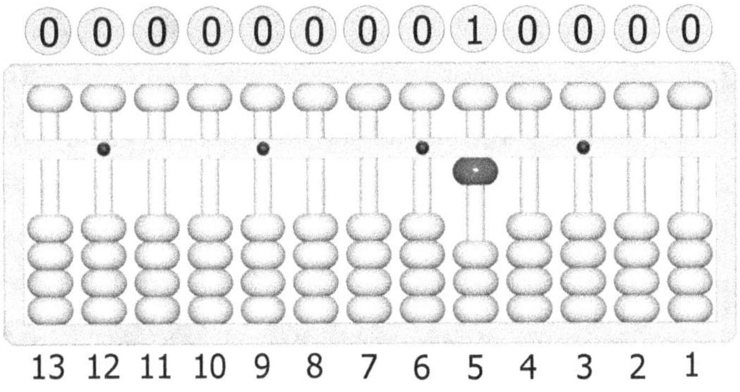

0 0 0 0 0 0 0 0 1 0 0 0 0

13 12 11 10 9 8 7 6 5 4 3 2 1

We will **register 100**

100 • Column 5, register 1 lower bead

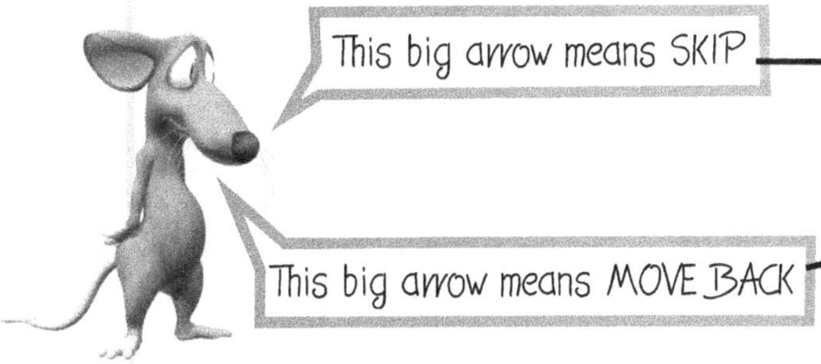

This big arrow means SKIP

This big arrow means MOVE BACK

The abacus reads 100

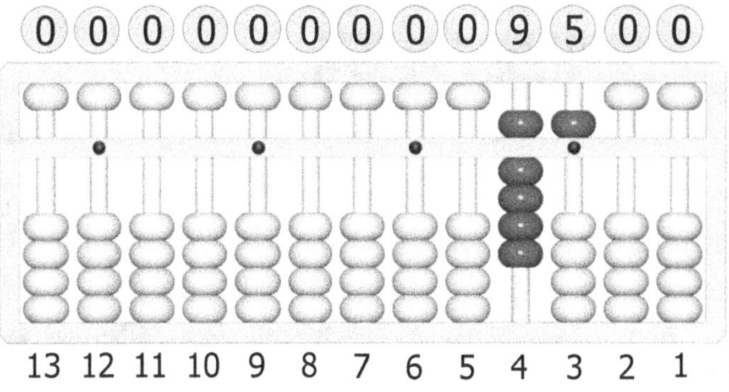

---

0 0 0 0 0 0 0 0 0 9 5 0 0

13 12 11 10 9 8 7 6 5 4 3 2 1

We will now **subtract 5**

There are not enough beads in column 3 to subtract 5, move to column 4, think **-5=-10+5**

← • Column 4, **SKIP** this column and
+90   register all beads

-100 • Column 5, unregister 1 lower bead

→ • **MOVE BACK** past the skipped column 4

+5 • Column 3, register 1 upper bead

The abacus result is 95

More skipped column examples

Example: 1000 - 1

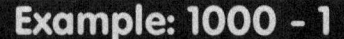

13 12 11 10 9 8 7 6 5 4 3 2 1

We will **register 1000**

1000 • Column 6, register 1 lower bead

Remember to move back past the skipped columns!

The abacus reads 1000

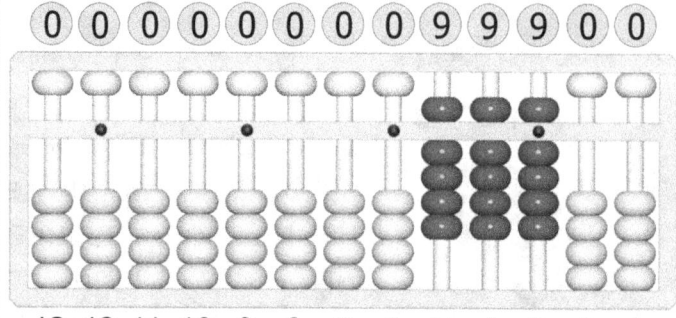

13 12 11 10 9 8 7 6 5 4 3 2 1

We will now **subtract 1**

There are not enough beads in column 3 to unregister 1, move to column 4, think **-1=-10+9**

← +90 • Column 4, **SKIP** this column and register all beads

← +900 • Column 5, **SKIP** this column and register all beads

-1000 • Column 6, unregister 1 lower bead

→ • **MOVE BACK** past the skipped

→ columns 5 and 4

+9 • Column 3, register 1 upper and 4 lower beads

The abacus result is 999

Example: 204 - 5

13 12 11 10 9 8 7 6 5 4 3 2 1

We will **register 204**

200 • Column 5, register 2 lower beads
     • Column 4, do nothing
4   • Column 3, register 4 lower beads

The abacus reads 204

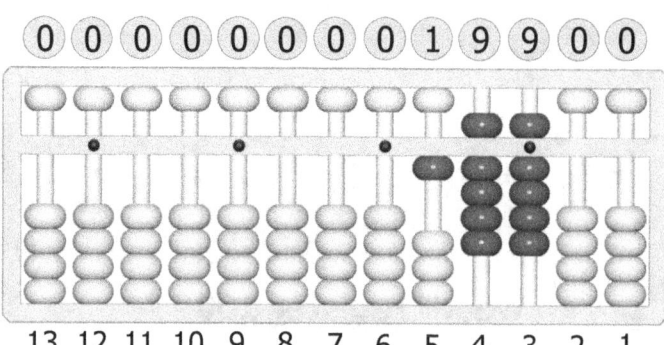

13 12 11 10 9 8 7 6 5 4 3 2 1

We will now **subtract 5**

There are not enough beads in column 3 to unregister 5, move to column 4, think **-5=-10+5**

 +90 • Column 4, **SKIP** this column and register all beads

-100 • Column 5, unregister 1 lower bead

 • **MOVE BACK** past the skipped column 4

+5 • Column 3, register 1 upper bead

The abacus result is 199

Subtraction of more than two numbers

Sometimes we have to subtract 3 or more numbers, here's how.

When we subtract many numbers on the abacus, just find the difference between the first two numbers, then subtract the next number to get the new difference.

Keep subtracting one number from the difference of the previous numbers until all the numbers have been subtracted.

Example: 998 - 332 - 151

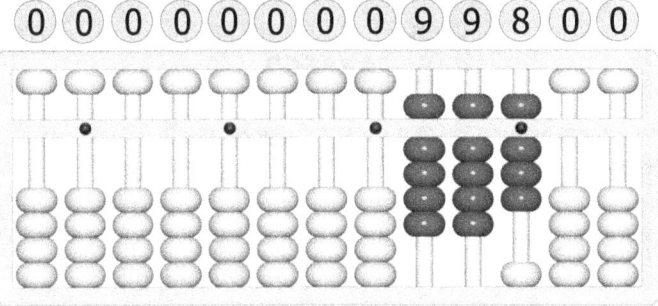

13 12 11 10 9 8 7 6 5 4 3 2 1

We will **register 998**

900 • Column 5, register 1 upper and 4 lower beads

90 • Column 4, register 1 upper and 4 lower beads

8 • Column 3, register 1 upper and 3 lower beads

The abacus reads 998

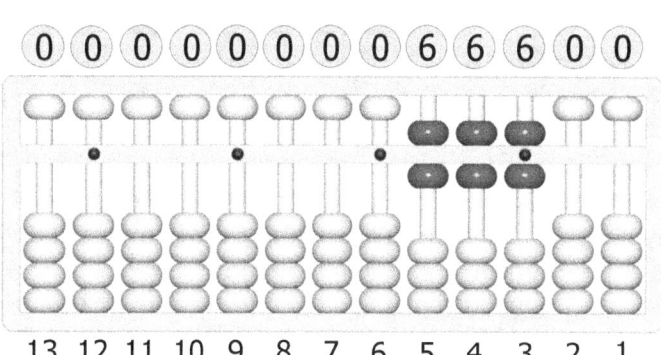

13 12 11 10 9 8 7 6 5 4 3 2 1

We will now **subtract 332 from 998**

-300 • Column 5, unregister 3 lower beads

-30 • Column 4, unregister 3 lower beads

-2 • Column 3, unregister 2 lower beads

The abacus now displays 666

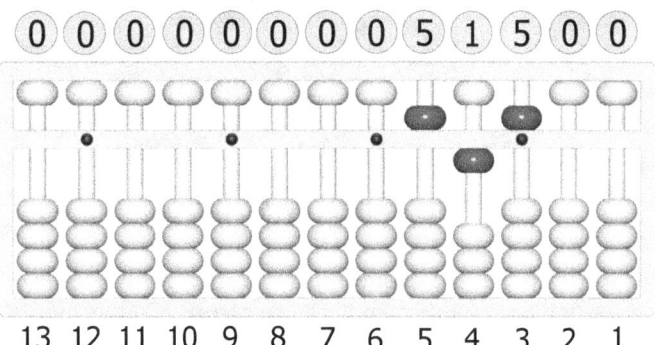

13 12 11 10 9 8 7 6 5 4 3 2 1

We will now **subtract 151 from 666**

-100 • Column 5, unregister 1 lower bead

-50 • Column 4, unregister 1 upper bead

-1 • Column 3, unregister 1 lower bead

The abacus result is 515

Subtraction of more than two numbers

Example: 43612424662 - 212330 - 1240

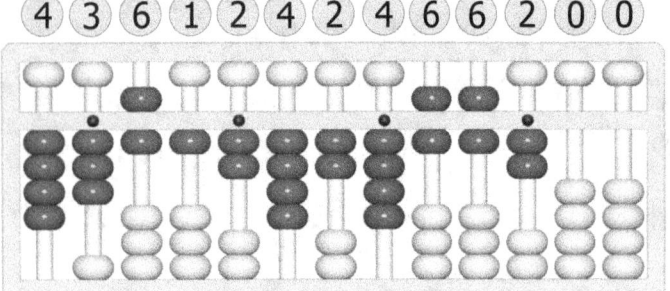

4	3	6	1	2	4	2	4	6	6	2	0	0

13 12 11 10 9 8 7 6 5 4 3 2 1

We will **register 43612424662**

40000000000	• Column 13, register 4 lower beads
3000000000	• Column 12, register 3 lower beads
600000000	• Column 11, register 1 upper and 1 lower bead
10000000	• Column 10, register 1 lower bead
2000000	• Column 9, register 2 lower beads
400000	• Column 8, register 4 lower beads
20000	• Column 7, register 2 lower beads
4000	• Column 6, register 4 lower beads
600	• Column 5, register 1 upper and 1 lower bead
60	• Column 4, register 1 upper and 1 lower bead
2	• Column 3, register 2 lower beads

The abacus reads 43612424662

4	3	6	1	2	2	1	2	3	3	2	0	0

13 12 11 10 9 8 7 6 5 4 3 2 1

We will now **subtract 212330 from 43612424662**

-200000	• Column 8, unregister 2 lower beads
-10000	• Column 7, unregister 1 lower bead
-2000	• Column 6, unregister 2 lower beads
-300	• Column 5, unregister 1 upper bead and register 2 lower beads
-30	• Column 4, unregister 1 upper bead and register 2 lower beads
	• Column 3, do nothing

The abacus now displays 43612212332

They were big numbers!

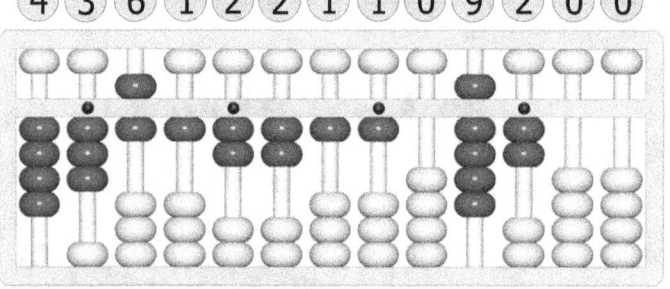

4	3	6	1	2	2	1	1	0	9	2	0	0

13 12 11 10 9 8 7 6 5 4 3 2 1

We will now **subtract 1240 from 43612212332**

| -1000 | • Column 6, unregister 1 lower bead |
| -200 | • Column 5, unregister 2 lower beads |

There are not enough beads in column 4 to unregister 4 more, move to column 5, think **-4=-10+6**

| -100 | • Column 5, unregister 1 lower bead |
| +60 | • Column 4, register 1 upper and 1 lower bead |

• Column 3, do nothing

The abacus result is 43612211092

TEST 4 - Subtraction

Try to subtract these numbers on your abacus.

Answers are on pages 64, 65 and 66.

 1

92 - 35

 6

635253 - 246062

 2

645 - 232

 7

5213634 - 6128

 3

814 - 8

 8

4536 - 224 - 12

 4

4342 - 2312

 9

440262 - 3201 - 435 - 6713

 5

66338 - 11223

 10

87654321 - 10000008 - 642107

Answers to test 4 (on page 63)

Here are the subtraction answers.
If you got any wrong, just have another go.

$92 - 35 = \mathbf{57}$

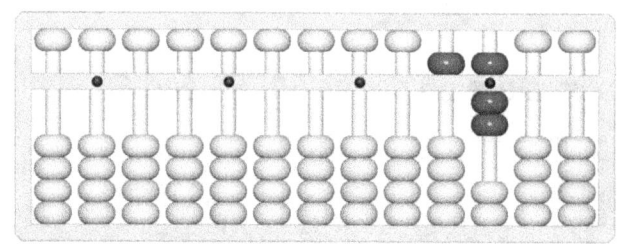

$645 - 232 = \mathbf{413}$

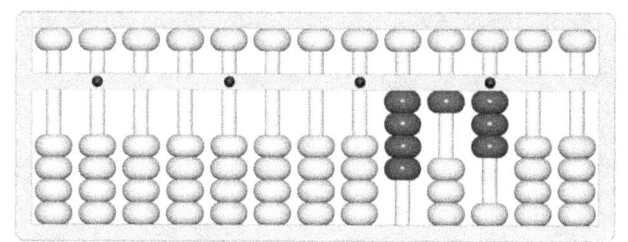

$814 - 8 = \mathbf{806}$

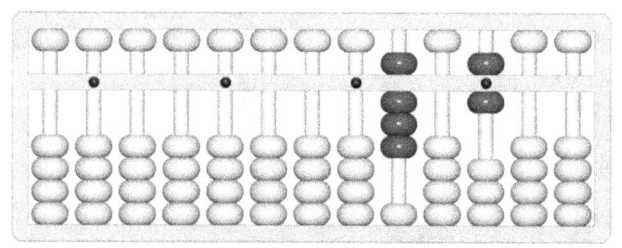

$4342 - 2312 = \mathbf{2030}$

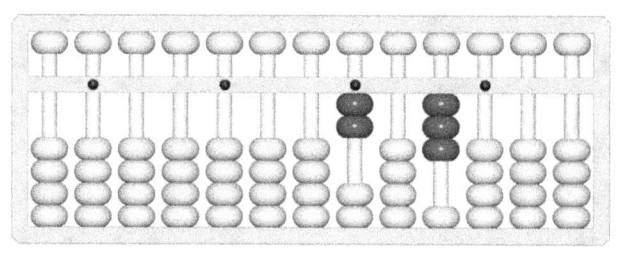

$66338 - 11223 = \mathbf{55115}$

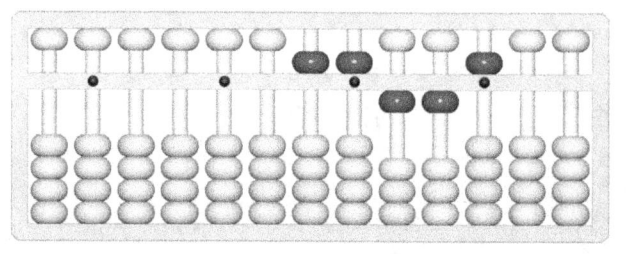

635253 - 246062 = **389191**

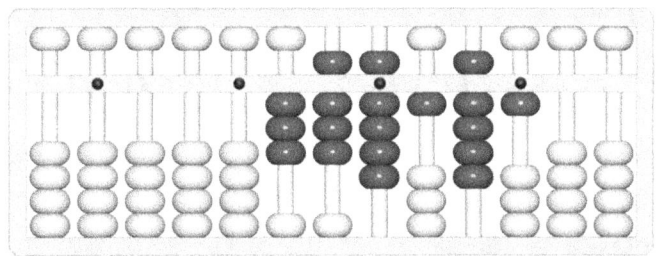

 5213634 - 6128 = **5207506**

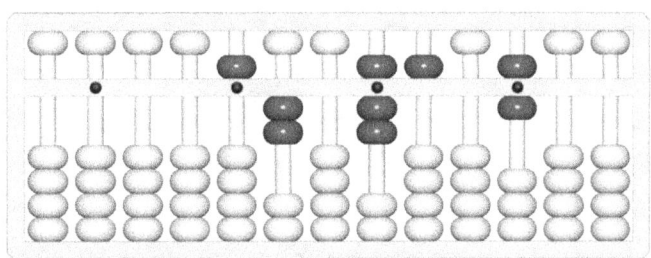

 4536 - 224 - 12 = **4300**

4536 - 224 = **4312**   4312 - 12 = **4300**

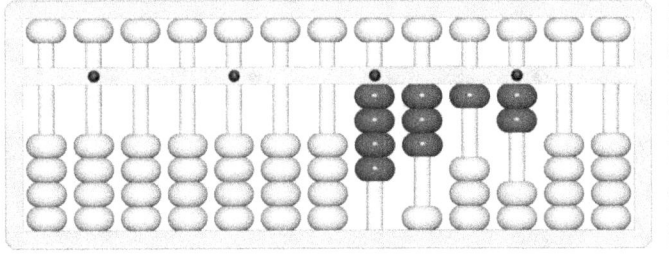 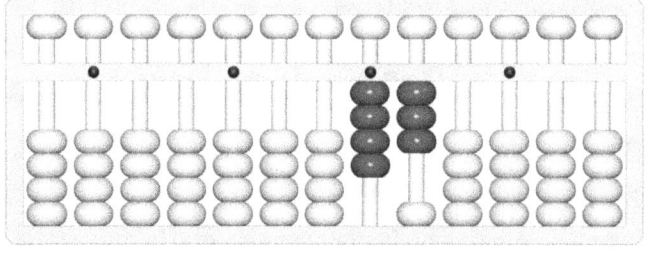

Answers to test 4 (on page 63)

 440262 - 3201 - 435 - 6713 = **429913**

440262 - 3201 = **437061**   437061 - 435 = **436626**

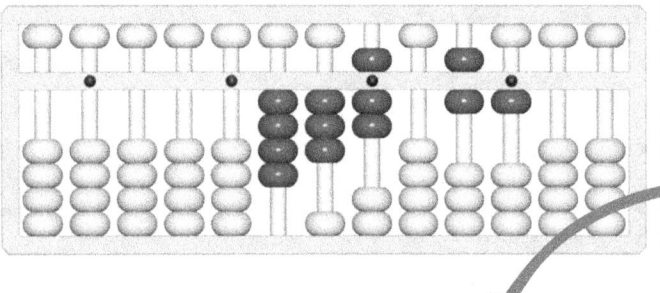

436626 - 6713 = **429913**

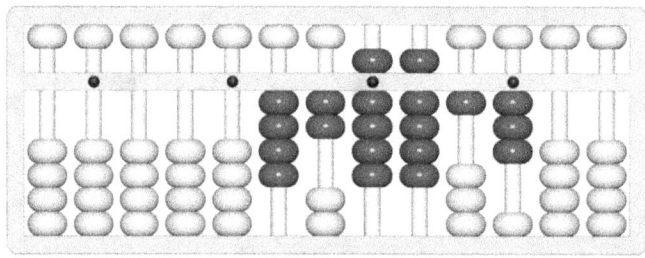

 87654321 - 10000008 - 642107 = **77012206**

87654321 - 10000008 = **77654313**   77654313 - 642107 = **77012206**

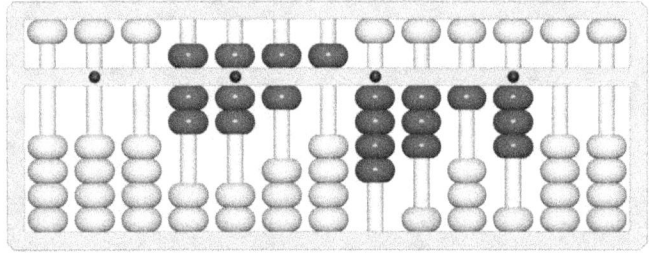 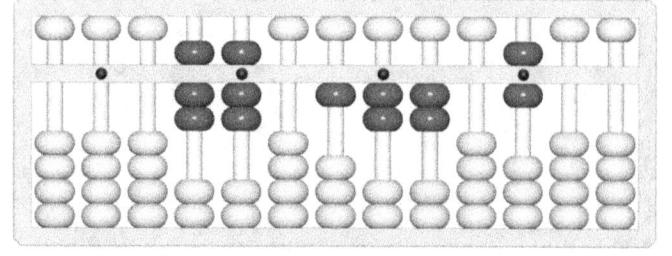

Subtraction of decimal numbers

Remember that a decimal number is a number that contains a decimal point, like 0.5 or 0.24

When we put a decimal number on the abacus, remember we use the dot on the beam to mark our ones column, so any number to the right of the dot (columns 1 & 2) will be our decimals.

Example: 46.23 - 12.1

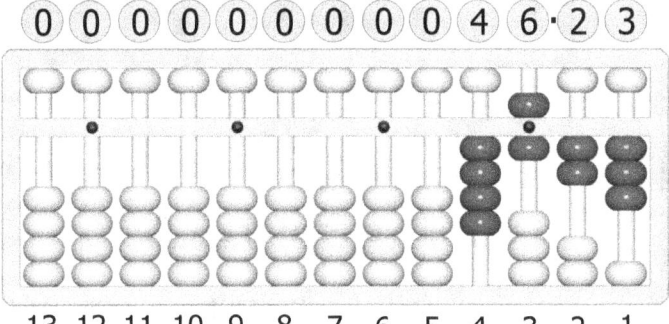

0 0 0 0 0 0 0 0 0 4 6·2 3
13 12 11 10 9 8 7 6 5 4 3 2 1

We will **register 46.23**

40	• Column 4, register 4 lower beads
6	• Column 3, register 1 upper and 1 lower bead
0.2	• Column 2, register 2 lower beads
0.03	• Column 1, register 3 lower beads

The abacus reads 46.23

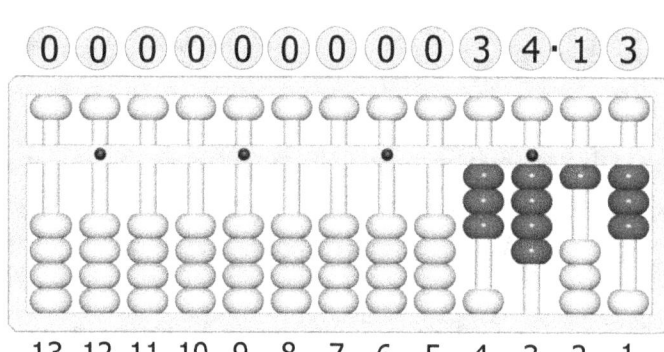

0 0 0 0 0 0 0 0 0 3 4·1 3
13 12 11 10 9 8 7 6 5 4 3 2 1

We will **subtract 12.1**

-10	• Column 4, unregister 1 lower bead
-2	• Column 3, unregister 1 upper bead and register 3 lower beads
-0.1	• Column 2, unregister 1 lower bead
	• Column 1, do nothing

The abacus result is 34.13

Example: 56.475 - 8.26

0 0 0 0 0 0 5 6·4 7 5 0 0
13 12 11 10 9 8 7 6 5 4 3 2 1

We will **register 56.475**

Use column 6 as the ones column

50	• Column 7, register 1 upper bead
6	• Column 6, register 1 upper and 1 lower bead
0.4	• Column 5, register 4 lower beads
0.07	• Column 4, register 1 upper and 2 lower beads
0.005	• Column 3, register 1 upper bead

The abacus reads 56.475

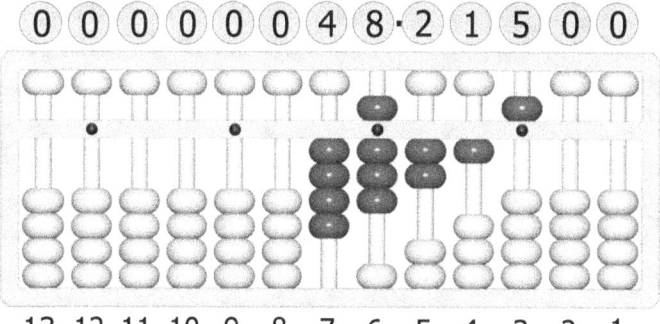

0 0 0 0 0 0 4 8·2 1 5 0 0
13 12 11 10 9 8 7 6 5 4 3 2 1

We will **subtract 8.26**

There are not enough beads in column 6, move to column 7, think **-8=-10+2**

-10	• Column 7, unregister 1 upper bead and register 4 lower beads
+2	• Column 6, register 2 lower beads
-0.2	• Column 5, unregister 2 lower beads
-0.06	• Column 4, unregister 1 upper and 1 lower bead
	• Column 3, do nothing

The abacus result is 48.215

More examples of subtraction of decimal numbers

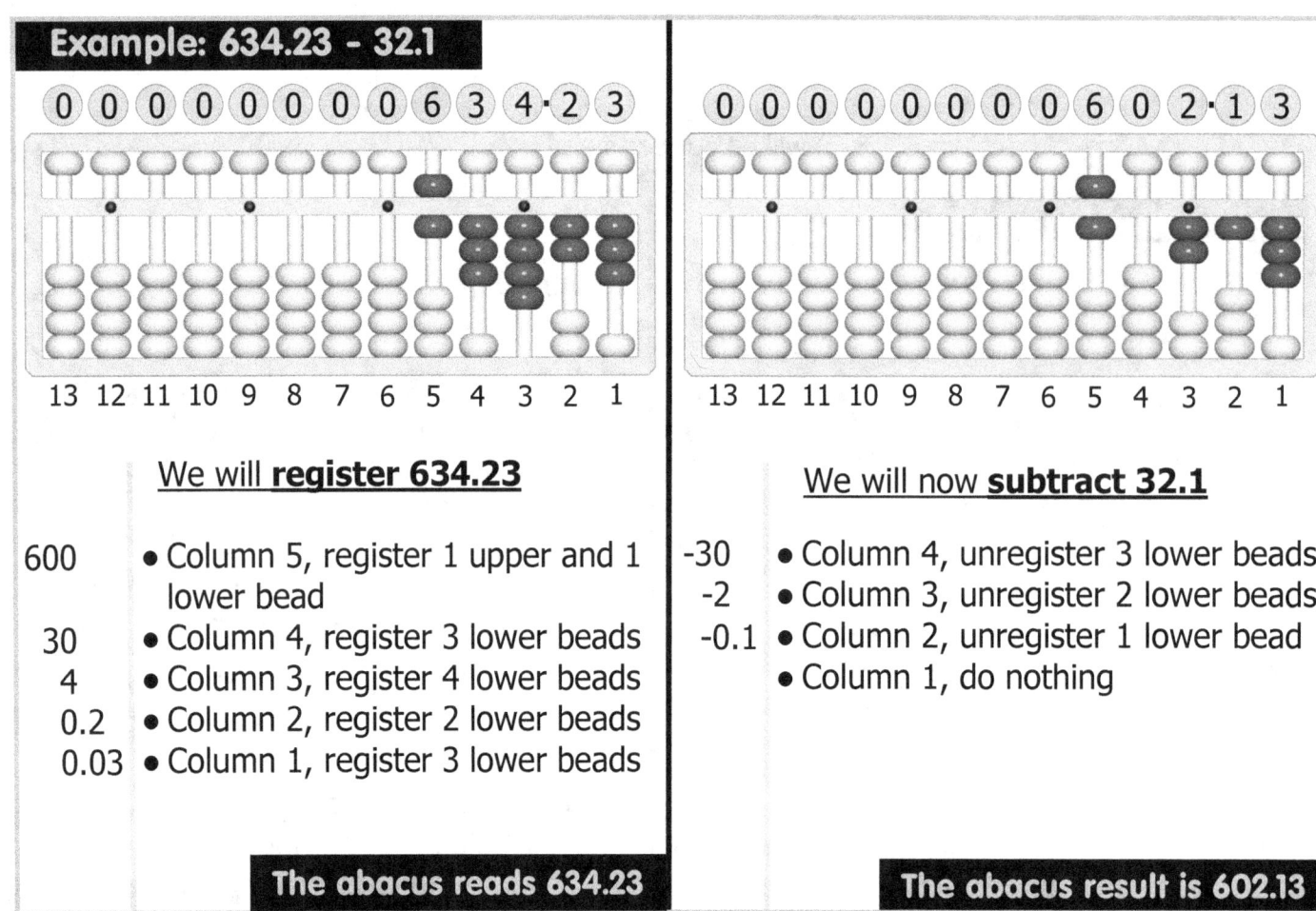

Example: 634.23 - 32.1

| 0 | 0 | 0 | 0 | 0 | 0 | 0 | 6 | 3 | 4·2 | 3 |

13 12 11 10 9 8 7 6 5 4 3 2 1

We will **register 634.23**

- 600 • Column 5, register 1 upper and 1 lower bead
- 30 • Column 4, register 3 lower beads
- 4 • Column 3, register 4 lower beads
- 0.2 • Column 2, register 2 lower beads
- 0.03 • Column 1, register 3 lower beads

The abacus reads 634.23

| 0 | 0 | 0 | 0 | 0 | 0 | 0 | 6 | 0 | 2·1 | 3 |

13 12 11 10 9 8 7 6 5 4 3 2 1

We will now **subtract 32.1**

- -30 • Column 4, unregister 3 lower beads
- -2 • Column 3, unregister 2 lower beads
- -0.1 • Column 2, unregister 1 lower bead
- • Column 1, do nothing

The abacus result is 602.13

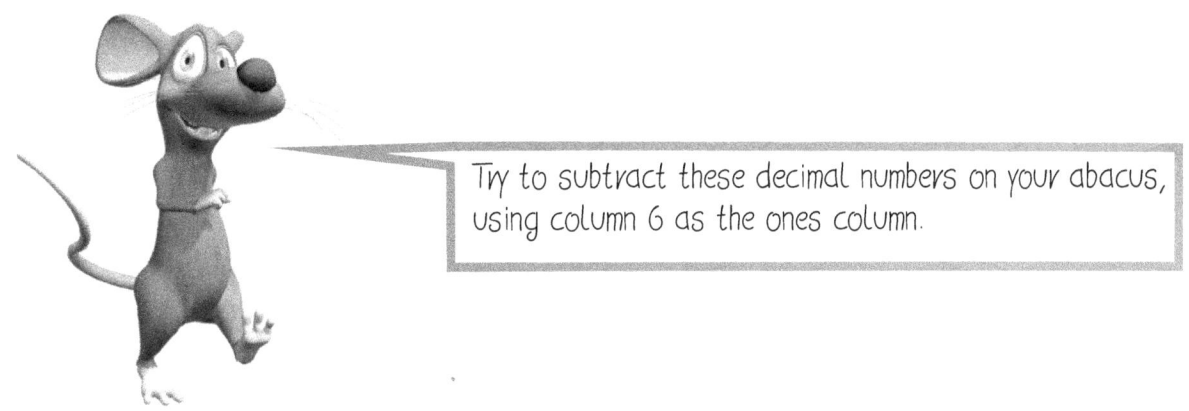

Try to subtract these decimal numbers on your abacus, using column 6 as the ones column.

Use column 6 as the ones column for these questions

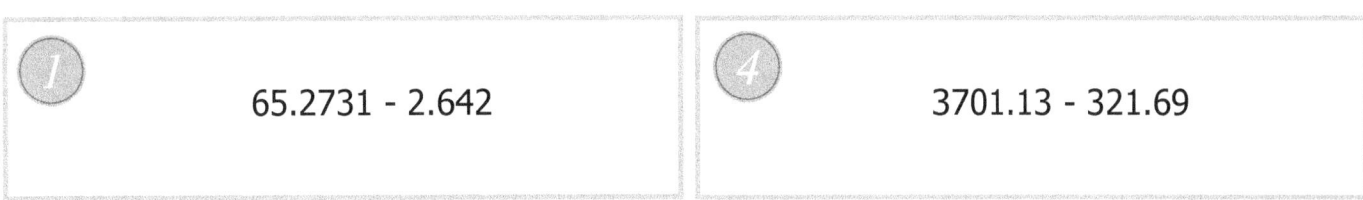

Answers are on page 70.

1	65.2731 - 2.642

4	3701.13 - 321.69

2	5.567 - 2.123

5	107.31646 - 92.64

3	54.039 - 12.36

70

Answers to test 5 (on page 69)

Here are the subtraction of decimal numbers answers.
If you got any wrong, just have another go.

65.2731 - 2.642 = **62.6311**

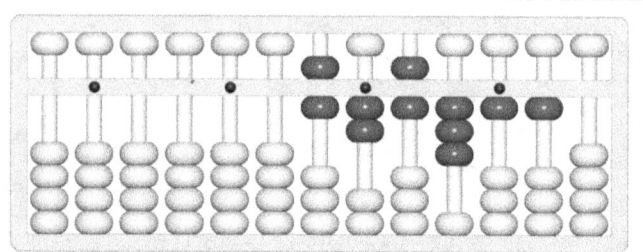

5.567 - 2.123 = **3.444**

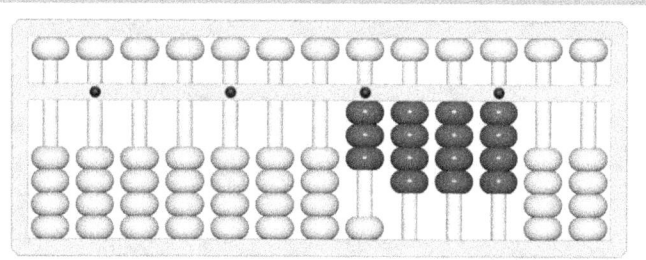

54.039 - 12.36 = **41.679**

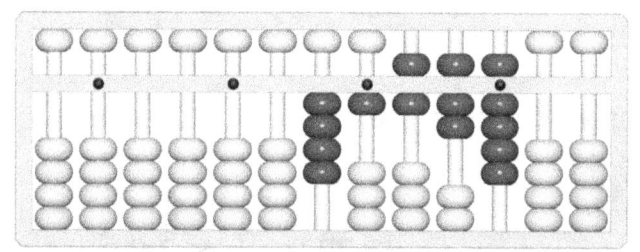

3701.13 - 321.69 = **3379.44**

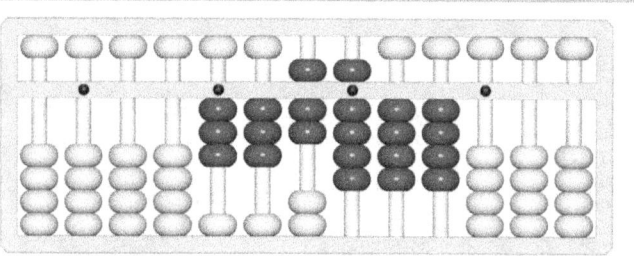

107.31646 - 92.64 = **14.67646**

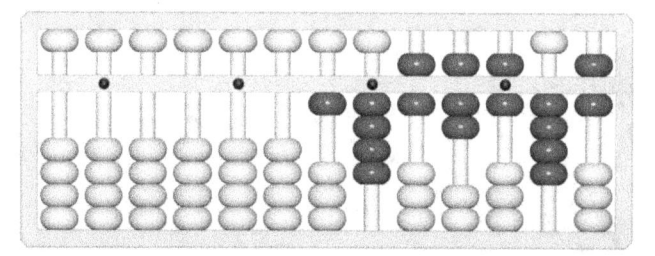

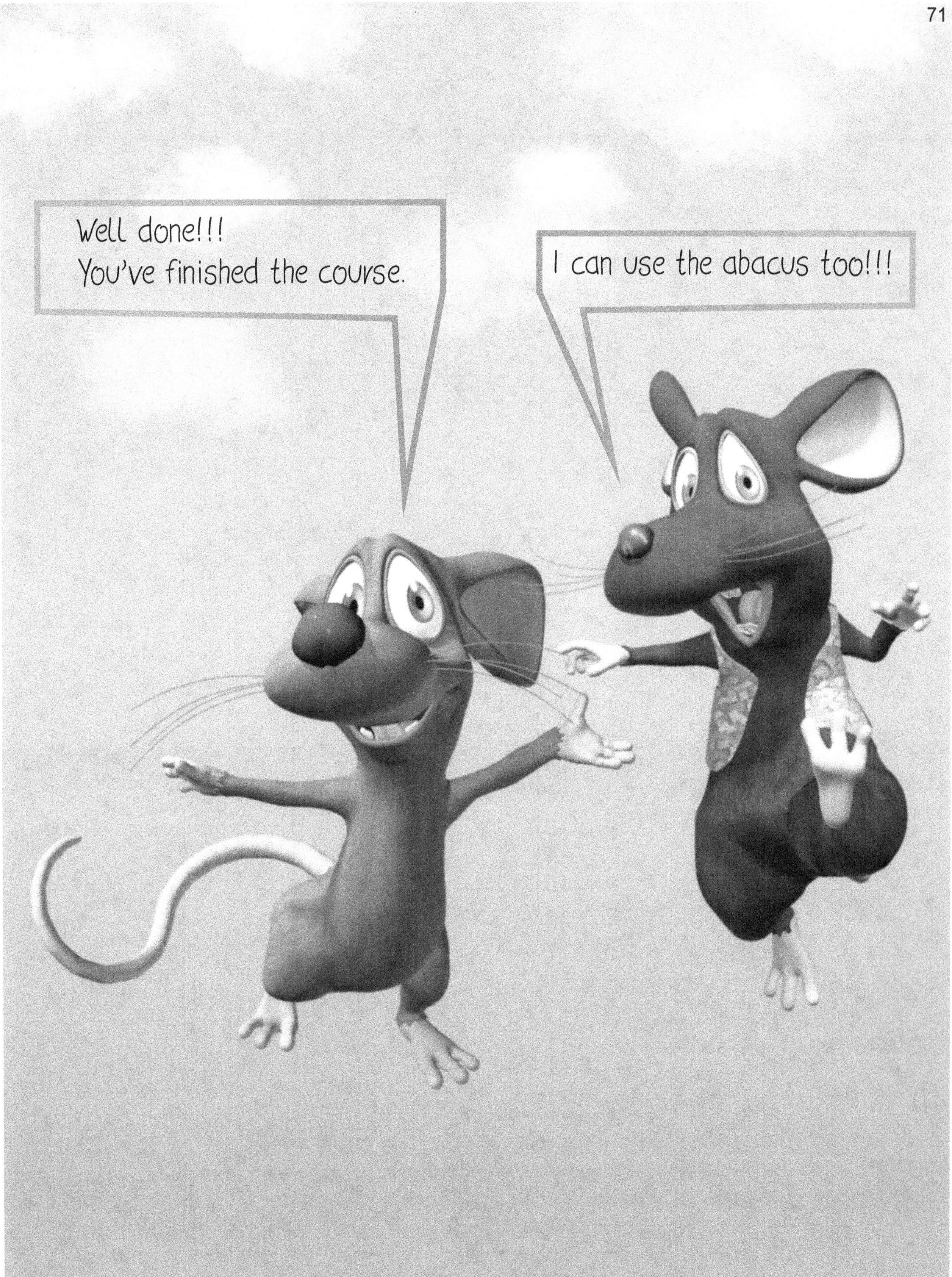
Well done!!!
You've finished the course.
I can use the abacus too!!!

MY NOTES

www.ingramcontent.com/pod-product-compliance
Lightning Source LLC
Chambersburg PA
CBHW081248180526
45170CB00007B/2343